The Transfer-Matrix Method in Electromagnetics and Optics

Synthesis Lectures on Electromagnetics

The Transfer-Matrix Method in Electromagnetics and Optics
Tom G. Mackay and Akhlesh Lakhtakia
2020

The Transfer-Matrix Method in Electromagnetics and Optics

Tom G. Mackay and Akhlesh Lakhtakia

ISBN: 978-3-031-00894-8   paperback
ISBN: 978-3-031-02022-3   ebook
ISBN: 978-3-031-00137-6   hardcover

DOI 10.1007/978-3-031-02022-3

A Publication in the Springer series
SYNTHESIS LECTURES ON ELECTROMAGNETICS

Lecture #1
Series ISSN
ISSN pending.

The Transfer-Matrix Method in Electromagnetics and Optics

Tom G. Mackay
University of Edinburgh

Akhlesh Lakhtakia
The Pennsylvania State University

SYNTHESIS LECTURES ON ELECTROMAGNETICS #1

ABSTRACT

The transfer-matrix method (TMM) in electromagnetics and optics is a powerful and convenient mathematical formalism for determining the planewave reflection and transmission characteristics of an infinitely extended slab of a linear material. While the TMM was introduced for a homogeneous uniaxial dielectric-magnetic material in the 1960s, and subsequently extended for multilayered slabs, it has more recently been developed for the most general linear materials, namely bianisotropic materials. By means of the rigorous coupled-wave approach, slabs that are periodically nonhomogeneous in the thickness direction can also be accommodated by the TMM. In this book an overview of the TMM is presented for the most general contexts as well as for some for illustrative simple cases. Key theoretical results are given; for derivations, the reader is referred to the references at the end of each chapter. Albums of numerical results are also provided, and the computer code used to generate these results are provided in an appendix.

KEYWORDS

transfer matrix, bianisotropy, periodic nonhomogeneity, rigorous coupled-wave approach, reflectance, transmittance, slab, matrix ordinary differential equation

Dedicated to the ethic of eco-responsibility championed by
Rachel Carson, Wangari Maathai, and Greta Thunberg

Contents

Preface

The transfer-matrix method (TMM) for linear electromagnetics and optics came into existence during the 1960s. Initiated in the 1966 doctoral thesis of Jean Billard for a homogeneous uniaxial dielectric-magnetic material, application of the TMM for multilayered slabs became popular after the publication of three papers of Dwight W. Berreman from 1970–1973. The rigorous coupled-wave approach widely used to solve grating problems emerged in the early 1980s from the papers of M. G. Moharam and Thomas K. Gaylord, its implementation requiring the TMM for electrically thin slabs.

In this book on the TMM, only key results are given, full details of derivations of these results being available in the references listed at the end of each chapter. The intended readership comprises graduate students and researchers, seeking a concise survey of the state-of-the-art about the TMM for electromagnetics and optics. A familiarity with undergraduate-level electromagnetic theory is assumed. SI units are used throughout.

Tom G. Mackay, Edinburgh, Scotland
Akhlesh Lakhtakia, University Park, PA, USA
January 2020

Acknowledgments

Every year, both of us learn more about electromagnetic theory and applications, not only through our own research but also by reviewing manuscripts, reading publications, attending seminars and conferences, and interacting with other researchers. We take joy in acknowledging our debt of gratitude to currently active colleagues worldwide as well as those who carried the beacons of science before us.

We especially thank Kevin Vynck of Institut d'Optique d'Aquitaine (Talence, France) for locating the Ph.D. thesis of Jean Billard, whose seminal contribution to the transfer-matrix method has remained obscure for five decades.

Akhlesh Lakhtakia thanks Faiz Ahmad, Tom H. Anderson, Francisco Chiadini, Benjamin J. Civiletti, Jhuma Dutta, Muhammad Faryad, Vincenzo Fiumara, Peter B. Monk, John A. Polo Jr., S. Anantha Ramakrishna, Antonio Scaglione, Mikhail V. Shuba, Manuel E. Solano, Vijayakumar C. Venugopal, and Fei Wang for research discussions that either directly or indirectly shaped this book. He also thanks the US National Science Foundation for two grants (DMR-1125591 and DMS-1619901), the Charles Godfrey Binder Endowment at The Pennsylvania State University for ongoing support of his research from 2006, and the Trustees of The Pennsylvania State University as well as the Otto Mønsted Foundation for enabling a sabbatical leave of absence at the Danish Technical University in Fall 2019. Tom Mackay acknowledges the support of EPSRC grant EP/S00033X/1.

We thank our families for their loving support and the staff of Morgan & Claypool for producing this book.

CHAPTER 1

Introduction

The transfer-matrix method (TMM) in electromagnetics and optics is a mathematically convenient formalism for determining the planewave reflection and transmission characteristics of an infinitely extended slab of a linear material. The direction of propagation and the polarization state of the incident plane wave can be arbitrary. This arbitrariness allows the TMM to be useful for the illumination of the slab by a finite source located at a finite distance from either of the two faces of the slab, because the time-harmonic fields radiated by that source can be expressed as an angular spectrum of plane waves [1]. The slab may be spatially homogeneous in the thickness direction or not. In the latter case, the slab may be continuously nonhomogeneous as for certain sculptured thin films [2] or the slab may be piecewise homogeneous in which case it is regarded as a multilayered slab [3, 4]. In a multilayered slab, the interface of any two adjacent constituent layers may be planar or periodically corrugated [5]. Finally, the materials in a slab may be of the most general linear type, i.e., bianisotropic materials [6].

Consider a bilayered slab occupying the region $0 < z < d$. Both constituent layers are homogeneous, have infinite extent along the x and y axes, and have finite thickness (along the z axis), as shown in Fig. 1.1(a). The interface of the two layers is planar and is parallel to both exposed faces of the bilayer. If a plane wave is considered incident on the bilayered slab, there must also exist a reflected plane wave and a transmitted plane wave. The TMM uses two 4×4 matrixes, one for each constituent layer in the bilayered slab, to relate the complex-valued amplitudes of the electric field phasor of the reflected and transmitted plane waves to the complex-valued amplitude of the electric field phasor of the incident plane wave. These 4×4 matrixes are called *transfer matrixes*.

Suppose the interface of the two constituent layers is periodically corrugated along the x axis and the direction of propagation of the incident plane wave lies wholly in the xz plane, as shown in Fig. 1.1(b). Then, the reflected electromagnetic field comprises an infinite number of distinct plane waves. These reflected plane waves are labeled $0, \pm1, \pm2, \ldots$. The reflected plane wave labeled 0 is called *specular*, the remaining ones being *nonspecular*. Only some of the nonspecular plane waves in the reflected field can transport energy an infinite distance from the bilayered slab. The transmitted electromagnetic field also comprises a specular plane wave (labeled 0) and an infinite number of nonspecular plane waves (labeled other than 0). Again, only some of the nonspecular plane waves in the transmitted field can transport energy an infinite distance from the bilayered slab. The TMM uses two $4(2M_t + 1) \times 4(2M_t + 1)$ matrixes, one for each constituent layer in the bilayered slab, to relate the complex-valued amplitudes of the

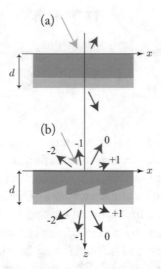

Figure 1.1: (a) Specular reflection and transmission of a plane wave incident on a bilayered slab wherein all interfaces are planar and parallel to each other. (b) Specular (labeled 0) and nonspecular (labeled other than 0) reflection and transmission when the internal interface in a bilayer is periodically corrugated along the x axis. The nonspecular modes are infinite in number, but only a few can propagate energy an infinite distance from the bilayered slab.

electric field phasors of the reflected and transmitted plane waves to the complex-valued amplitude of the electric field phasor of the incident plane wave, with the integer $M_t > 0$ being sufficiently large.

When the interface of the two constituent layers is periodically corrugated along both the x and y axes, the TMM uses two $4(2\tau_t + 1) \times 4(2\tau_t + 1)$ matrixes, one for each layer in the bilayered slab, where $\tau_t = M_t(N_t + 1) + N_t(M_t + 1)$ and the integers $M_t > 0$ and $N_t > 0$ are sufficiently large.

1.1   BRIEF HISTORY OF TMM

Before the advent of the TMM, recursive schemes had been formulated and implemented for multilayered slabs comprising planar layers of isotropic dielectric materials [7–10]. If the incident plane wave is linearly polarized, the reflected and transmitted plane waves then have the same polarization state as the incident plane wave. These schemes can be extended to incorporate periodically corrugated interfaces [11, 12], but become cumbersome [13] when the reflected and transmitted plane waves can have both co-polarized and cross-polarized components. That happens when: (i) materials more complicated than isotropic dielectric-magnetic materials [6, 14] are involved, and/or (ii) the direction of propagation of the incident plane wave has a nonzero component along the y axis and an interface is periodically corrugated along the x axis, and/or

(iii) when the interface is periodically corrugated along both the x and y axes [5, 15, 16]. The TMM is very convenient when such issues must be tackled.

When a linearly polarized plane wave is incident on a slab of an isotropic dielectric material, the electromagnetic fields induced inside it can be decomposed into two plane waves, one propagating toward one face of the slab and the other propagating toward the other face. Both of these plane waves have the same polarization state as the incident plane wave. The amplitudes of the electric field phasors of the induced plane waves can be used to formulate a 2×2 matrix, which can be used to develop a matrix formulation [17–21] for multilayered slabs with planar interfaces. This matrix formulation can be extended to encompass anisotropic dielectric materials [12], but the extension is inelegant and definitely cumbersome.

For any time-harmonic electromagnetic field in a source-free region occupied by any homogeneous isotropic dielectric-magnetic medium, the derivative $(\partial/\partial z)\left[\left(\underline{u}_x\underline{u}_x + \underline{u}_y\underline{u}_y\right) \cdot \underline{E}(\underline{r}, \omega)\right]$ can be written in terms of the x- and y-directed components of $\underline{H}(\underline{r}, \omega)$ and their derivatives with respect to x and y, where $\underline{r} = x\underline{u}_x + y\underline{u}_y + z\underline{u}_z$ is the position vector with $\{\underline{u}_x, \underline{u}_y, \underline{u}_z\}$ as the triad of Cartesian unit vectors, and ω is the angular frequency. Likewise, the derivative $(\partial/\partial z)\left[\left(\underline{u}_x\underline{u}_x + \underline{u}_y\underline{u}_y\right) \cdot \underline{H}(\underline{r}, \omega)\right]$ can be written in terms of the x- and y-directed components of $\underline{E}(\underline{r}, \omega)$ and their derivatives with respect to x and y [22]. These relationships can be extended to source-free regions occupied by homogeneous bianisotropic materials [23]. Accordingly, if the field phasors are expressed as

$$\left.\begin{aligned} \underline{E}(\underline{r}, \omega) &= \underline{e}(z, \omega) \, \exp\left[i q(x \cos \psi + y \sin \psi)\right] \\ \underline{H}(\underline{r}, \omega) &= \underline{h}(z, \omega) \, \exp\left[i q(x \cos \psi + y \sin \psi)\right] \end{aligned}\right\}, \tag{1.1}$$

with auxiliary phasors

$$\left.\begin{aligned} \underline{e}(z, \omega) &= e_x(z, \omega)\underline{u}_x + e_y(z, \omega)\underline{u}_y + e_z(z, \omega)\underline{u}_z \\ \underline{h}(z, \omega) &= h_x(z, \omega)\underline{u}_x + h_y(z, \omega)\underline{u}_y + h_z(z, \omega)\underline{u}_z \end{aligned}\right\}, \tag{1.2}$$

the 4×4 matrix ordinary differential equation

$$\frac{d}{dz}\left[\underline{f}(z, \omega)\right] = i \left[\underline{\underline{P}}(\omega)\right] \cdot \left[\underline{f}(z, \omega)\right] \tag{1.3}$$

is obtained. Herein, $q(\underline{u}_x \cos \psi + \underline{u}_y \sin \psi)$ is the transverse wave vector with $\psi \in [0, 2\pi)$ as an angle, the 4-column vector

$$\left[\underline{f}(z, \omega)\right] = \begin{bmatrix} e_x(z, \omega) \\ e_y(z, \omega) \\ h_x(z, \omega) \\ h_y(z, \omega) \end{bmatrix}, \tag{1.4}$$

while the 4×4 matrix $\left[\underline{\underline{P}}(\omega)\right]$ is specified in (2.49).

Equation (1.3) is the bedrock of the TMM. It was formulated and solved for propagation in a uniaxial dielectric-magnetic material by Billard in 1966 [24], although the mathematical foundation had been laid during the previous decade [25]. Shortly thereafter, Teitler and Henvis [26] formulated and solved (1.3) for a slab of an anisotropic dielectric material, but then they reverted to the 2×2-matrix formalism of Abelès that had originated two decades earlier [20]. However, Berreman persevered with 4×4 matrixes, first for slabs of cholesteric liquid crystals [27] and then for slabs of homogeneous bianisotropic materials [28].

Early attempts to apply the TMM to a slab composed of a continuously nonhomogeneous material must be viewed with caution. Although Berreman [27] correctly applied a numerical technique to solve (1.3) for a cholesteric liquid crystal of finite thickness, Equation (6) in Ref. 29 is incorrect [30]. A piecewise-uniform approximation provides a convenient path to handle slabs of continuously nonhomogeneous materials [31, 32], but other numerical techniques also exist [33, 34]. The TMM has also been developed to accommodate anisotropic [15] and bianisotropic [5, 16] layers with periodically corrugated interfaces.

1.2   APPLICATIONS OF TMM

A commonplace experimental configuration, i.e., a light beam incident on a slab, provides the backdrop for the TMM. A beam of large width in relation to the wavelength can be approximated by a plane wave [1]. Accordingly, the TMM is useful for a host of practical applications.

If the slab is a single layer of a homogeneous or nonhomogeneous material, the TMM can be harnessed with ellipsometry measurements [35, 36] to characterize the optical properties of the slab material. Furthermore, the TMM can be exploited to aid the design of optical components such as Bragg mirrors, antireflection coatings, waveplates, and polarization converters [36–38].

Another major area of TMM application for multilayered slabs lies in the analysis of surface waves that are guided by the planar interface of two dissimilar materials [5, 39]. For example, in the case of spatially homogeneous constituent layers, Dyakonov surface waves [40–42] can be excited at the interface of two dielectric constituent layers provided that one of them is anisotropic, while surface-plasmon-polariton waves [43, 44] can be excited at the interface of two isotropic constituent layers provided that one of them is metallic. If one (or more) of the constituent layers is periodically nonhomogeneous in the thickness direction, then Tamm surface waves [45, 46] can be excited at the interface of two isotropic constituent layers or Dyakonov–Tamm surface waves [47, 48] can be excited provided that one of the constituent layers is anisotropic. Composite guided waves excited in a multilayered slab can also be identified using the TMM [49].

If the interfaces of adjacent constituent layers in a multilayered slab are planar, the slab provides an appropriate model for the analysis of prism-coupled excitation of surface waves, as arises in the Turbadar–Kretschmann–Raether [50–52] and Turbadar–Otto [53] configurations. On the other hand, if an interface of two adjacent constituent layers is periodically corrugated, then the TMM can be harnessed to analyze surface waves that are excited in the grating-coupled configuration [43, 54, 55].

The versatility of the TMM extends to the incorporation of a layer of a topological insulator [56, 57] in a slab through suitable jump conditions across the two faces of that layer [58]. Also, infinitely extended graphene [59] can be incorporated through a jump condition across it [60].

1.3   OVERVIEW

The TMM is presented in full generality in this book. Chapter 2 provides the prerequisite electromagnetic theory concerning linear materials; bianisotropic materials and specializations thereof are described, and the 4×4-matrix formulation for planewave propagation in such materials is introduced. The TMM for a slab with adjacent constituent layers that have planar interfaces is presented in Chapter 3. The material of each constituent layer is spatially homogeneous and bianisotropic in general. By means of a piecewise-uniform approximation, the same formalism can be applied to a bianisotropic slab which is spatially nonhomogeneous in the thickness direction. The TMM formalism of Chapter 3 is extended in Chapter 4 wherein a multilayered slab containing two adjacent constituent layers with a doubly periodic interface is considered; as in Chapter 3, the constituent layers in Chapter 4 are spatially homogeneous and bianisotropic. In order to highlight the key features of the TMM for the simplest scenario, the TMM is given in Chapter 5 for a multilayered slab made of isotropic dielectric materials, and the constituent layers can have either planar interfaces or singly periodic interfaces. Chapter 6 comprises some closing remarks. A short overview of 3×3 dyadics is provided in Appendix A; for further details, readers are referred elsewhere [61]. Mathematica™ codes used to generate the numerical results presented in Chapters 3 and 5 are provided in Appendix B.

In the notation adopted, 3-vectors are underlined once while 3×3 dyadics [61] are double underlined. Matrixes are doubled underlined and enclosed in square brackets; and n-vectors where $n > 3$ are underlined once and enclosed in square brackets. The identity 3×3 dyadic is denoted by $\underline{\underline{I}} = \underline{u}_x \underline{u}_x + \underline{u}_y \underline{u}_y + \underline{u}_z \underline{u}_z$ and the null 3×3 dyadic by $\underline{\underline{0}}$. The real and imaginary parts of complex quantities are delivered by the operators $\mathrm{Re}\{\bullet\}$ and $\mathrm{Im}\{\bullet\}$, respectively. The complex conjugate of a complex-valued scalar ζ is denoted by ζ^*. The symbols ε_0 and μ_0 represent the permittivity and permeability of free space, respectively. The free-space wavenumber is denoted by $k_0 = \omega\sqrt{\varepsilon_0\mu_0}$, the wavelength in free space by $\lambda_0 = 2\pi/k_0$, and the intrinsic impedance of free space by $\eta_0 = \sqrt{\mu_0/\varepsilon_0}$. SI units are adopted throughout.

1.4   REFERENCES

[1] Clemmow, P. C. 1966. *The Plane Wave Spectrum Representation of Electromagnetic Fields* (New York, Pergamon Press). DOI: 10.1109/9780470546598. 1, 4

[2] Lakhtakia, A. and Messier, R. 2005. *Sculptured Thin Films: Nanoengineered Morphology and Optics* (Bellingham, WA, SPIE Press). DOI: 10.1117/3.585322. 1

[3] Huxley, A. F. 1968. A theoretical treatment of the reflexion of light by multilayer structures, *J. Exp. Biol.*, 48:227–245. 1

[4] Yeh, P. 1980. Optics of anisotropic layered media: A new 4 × 4 matrix algebra, *Surf. Sci.*, 96:41–53. DOI: 10.1016/0039-6028(80)90293-9. 1

[5] Polo, J. A., Jr., Mackay, T. G., and Lakhtakia, A. 2013. *Electromagnetic Surface Waves: A Modern Perspective* (Waltham, MA, Elsevier). 1, 3, 4

[6] Mackay, T. G. and Lakhtakia, A. 2019. *Electromagnetic Anisotropy and Bianisotropy: A Field Guide*, 2nd ed. (Singapore, World Scientific). DOI: 10.1142/11351. 1, 2

[7] Rouard, P. 1937. Études des propriétés optiques des lames métalliques très minces, *Ann. Phys.*, 11:291–384, Paris. DOI: 10.1051/anphys/193711070291. 2

[8] Vašiček, A. 1947. The reflection of light from glass with double and multiple films, *J. Opt. Soc. Am.*, 37:954–964. DOI: 10.1364/josa.37.000623. 2

[9] Crook, A. W. 1948. The reflection and transmission of light by any system of parallel isotropic films, *J. Opt. Soc. Am.*, 38:623–634. DOI: 10.1364/josa.38.000954. 2

[10] Parratt, L. G. 1954. Surface studies of solids by total reflection of X-rays, *Phys. Rev.*, 95:359–369. DOI: 10.1103/physrev.95.359. 2

[11] Cain, W. N., Varadan, V. K., Varadan, V. V., and Lakhtakia, A. 1986. Reflection and transmission characteristics of a slab with periodically varying surfaces, *IEEE Trans. Antennas Propagat.*, 34:1159–1163. DOI: 10.1109/tap.1986.1143949. 2

[12] Abdulhalim, I. 1999. 2 × 2 Matrix summation method for multiple reflections and transmissions in a biaxial slab between two anisotropic media, *Opt. Commun.*, 163:9–14. DOI: 10.1016/s0030-4018(99)00129-7. 2, 3

[13] Heavens, O. S. 1960. Optical properties of thin films, *Rep. Prog. Phys.*, 23:1–65. DOI: 10.1088/0034-4885/23/1/301. 2

[14] Faryad, M. and Lakhtakia, A. 2018. *Infinite-Space Dyadic Green Functions in Electromagnetism* (San Rafael, CA, Morgan & Claypool), IOP Concise Physics. DOI: 10.1088/978-1-6817-4557-2. 2

[15] Glytsis, E. N. and Gaylord, T. K. 1987. Rigorous three-dimensional coupled-wave diffraction analysis of single and cascaded anisotropic gratings, *J. Opt. Soc. Am. A*, 4:2061–2080. DOI: 10.1364/josaa.4.002061. 3, 4

[16] Onishi, M., Crabtree, K., and Chipman, R. C. 2011. Formulation of rigorous coupled-wave theory for gratings in bianisotropic media, *J. Opt. Soc. Am. A*, 28:1747–1758. DOI: 10.1364/josaa.28.001747. 3, 4

[17] Weinstein, W. 1947. The reflectivity and transmissivity of multiple thin coatings, *J. Opt. Soc. Am.*, 37:576–581. DOI: 10.1364/josa.37.000576. 3

[18] Herpin, A. 1947. Calcul du pouvoir réflecteur d'un système stratifié quelconque, *C. R. Hebdo. Séances l'Acad. Sci.*, 225:182–183, Paris. 3

[19] Muchmore, R. B. 1948. Optimum band width for two layer anti-reflection films, *J. Opt. Soc. Am.*, 38:20–26. DOI: 10.1364/josa.38.000020. 3

[20] Abelès, F. 1950. Recherches sur la propagation des ondes électromagnétiques sinusoïdales dans les milieux stratifiés. Application aux couches minces (1re partie), *Ann. Phys.*, 5:596–640, Paris. DOI: 10.1051/anphys/195012050706. 3, 4

[21] Abelès, F. 1957. Optical properties of thin absorbing films, *J. Opt. Soc. Am.*, 47:473–482. DOI: 10.1364/josa.47.000473. 3

[22] Marcuvitz, N. and Schwinger, J. S. 1951. On the representation of the electric and magnetic fields produced by currents and discontinuities in wave guides, I, *J. Appl. Phys.*, 22:806–819. DOI: 10.1063/1.1700052. 3

[23] Weiglhofer, W. 1987. Scalarization of Maxwell's equations in general inhomogeneous bianisotropic media, *IEE Proc. H*, 134:357–360. DOI: 10.1049/ip-h-2.1987.0070. 3

[24] Billard, J. 1966. *Contribution a l'etude de la propagation des ondes electromagnetiques planes dans certains milieux materiels (2ème these)*, Ph.D. Dissertation (Université de Paris 6, France), pages 175–178. 4

[25] Keller, H. B. and Keller, J. B. 1962. Exponential-like solutions of systems of linear ordinary differential equations, *J. Soc. Indust. Appl. Math.*, 10(2):246–259. DOI: 10.1137/0110019. 4

[26] Teitler, S. and Henvis, B. W. 1970. Refraction in stratified, anisotropic media, *J. Opt. Soc. Am.*, 60:830–834. DOI: 10.1364/josa.60.000830. 4

[27] Berreman, D. W. and Scheffer, T. J. 1970. Bragg reflection of light from single-domain cholesteric liquid-crystal films, *Phys. Rev. Lett.*, 25:577–581. DOI: 10.1103/phys-revlett.25.902.4. 4

[28] Berreman, D. W. 1972. Optics in stratified and anisotropic media: 4×4-matrix formulation, *J. Opt. Soc. Am.*, 62:502–510. DOI: 10.1364/josa.62.000502. 4

[29] Berreman, D. W. 1973. Optics in smoothly varying anisotropic planar structures: Application to liquid-crystal twist cells, *J. Opt. Soc. Am.*, 63:1374–1380. DOI: 10.1364/josa.63.001374. 4

[30] Lakhtakia, A. 2003. Comment on 'Analytical solution of non-homogeneous anisotropic wave equations based on differential transfer matrices,' *J. Opt. A: Pure Appl. Opt.*, 5:432–433. DOI: 10.1088/1464-4258/5/4/401. 4

[31] Abdulhalim, I., Benguigui, L., and Weil, R. 1985. Selective reflection by helicoidal liquid crystals. Results of an exact calculation using the 4×4 characteristic matrix method, *J. Phys.*, 46:815–825, Paris. DOI: 10.1051/jphys:01985004605081500. 4

[32] Venugopal, V. C. and Lakhtakia, A. 1998. Dielectric sculptured nematic thin films for rugate-like filters, *Opt. Commun.*, 149:217–222. DOI: 10.1016/s0030-4018(98)00029-7. 4

[33] Venugopal, V. C. and Lakhtakia, A. 2000. Electromagnetic plane-wave response characteristics of non-axially excited slabs of dielectric thin-film helicoidal bianisotropic mediums, *Proc. R. Soc. Lond. A*, 456:125–161. DOI: 10.1098/rspa.2000.0511. 4

[34] Polo, J. A., Jr., and Lakhtakia, A. 2002. Numerical implementation of exact analytical solution for oblique propagation in a cholesteric liquid crystal, *Microw. Opt. Technol. Lett.*, 35:397–400. DOI: 10.1002/mop.10618. 4

[35] Ward, L. 2000. *The Optical Constants of Bulk Materials and Films*, 2nd ed. (Bristol, UK, Institute of Physics). 4

[36] Hodgkinson, I. J. and Wu, Q. h. 1997. *Birefringent Thin Films and Polarizing Elements* (Singapore, World Scientific). DOI: 10.1142/3324. 4

[37] MacLeod, H. A. 2001. *Thin-Film Optical Filters*, 3rd ed. (Boca Raton, FL, CRC Press). DOI: 10.1201/9781420033236. 4

[38] Baumeister, P. W. 2004. *Optical Coating Technology* (Bellingham, WA, SPIE Press). DOI: 10.1117/3.548071. 4

[39] Boardman, A. D., Ed., 1982. *Electromagnetic Surface Modes* (Chicester, UK, Wiley). 4

[40] Marchevskiĭ, F. N., Strizhevskiĭ, V. L., and Strizhevskiĭ, S. V. 1984. Singular electromagnetic waves in bounded anisotropic media, *Sov. Phys. Solid State*, 26:911–912. 4

[41] D'yakonov, M. I. 1988. New type of electromagnetic wave propagating at an interface, *Sov. Phys. JETP*, 67:714–716. 4

[42] Takayama, O., Crasovan, L., Artigas, D., and Torner, L. 2009. Observation of Dyakonov surface waves, *Phys. Rev. Lett.*, 102:043903. DOI: 10.1103/physrevlett.102.043903. 4

[43] Homola, J., Ed., 2006. *Surface Plasmon Resonance Based Sensors* (Heidelberg, Germany, Springer). DOI: 10.1007/b100321. 4, 5

[44] Pitarke, J. M., Silkin, V. M., Chulkov, E. V., and Echenique, P. M. 2007. Theory of surface plasmons and surface-plasmon polaritons, *Rep. Prog. Phys.*, 70:1–87. DOI: 10.1088/0034-4885/70/1/r01. 4

[45] Yeh, P., Yariv, A., and Hong, C.-S. 1977. Electromagnetic propagation in periodic stratified media. I. General theory, *J. Opt. Soc. Am.*, 67:423–438. DOI: 10.1364/josa.67.000423. 4

[46] Martorell, J., Sprung, D. W. L., and Morozov, G. V. 2006. Surface TE waves on 1D photonic crystals, *J. Opt. A: Pure Appl. Opt.*, 8:630–638. DOI: 10.1088/1464-4258/8/8/003. 4

[47] Lakhtakia, A. and Polo, J. A., Jr. 2007. Dyakonov–Tamm wave at the planar interface of a chiral sculptured thin film and an isotropic dielectric material, *J. Eur. Opt. Soc.—Rapid Pub.*, 2:07021. DOI: 10.2971/jeos.2007.07021. 4

[48] Pulsifer, D. P., Faryad, M., and Lakhtakia, A. 2013. Observation of the Dyakonov–Tamm wave, *Phys. Rev. Lett.*, 111:243902. DOI: 10.1103/physrevlett.111.243902. 4

[49] Chiadini, F., Fiumara, V., Scaglione, A., and Lakhtakia, A. 2016. Compound guided waves that mix characteristics of surface-plasmon-polariton, Tamm, Dyakonov–Tamm, and Uller–Zenneck waves, *J. Opt. Soc. Am. B*, 33:1197–1206. DOI: 10.1364/josab.33.001197. 4

[50] Turbadar, T. 1959. Complete absorption of light by thin metal films, *Proc. Phys. Soc.*, 73:40–44. DOI: 10.1088/0370-1328/73/1/307. 4

[51] Turbadar, T. 1964. Complete absorption of plane polarized light by thin metal films, *Opt. Acta*, 11:207–210. DOI: 10.1080/713817875. 4

[52] Kretschmann, E. and Raether, H. 1968. Radiative decay of nonradiative surface plasmons excited by light, *Z. Naturforsch. A*, 23:2135–2136. DOI: 10.1515/zna-1968-1247. 4

[53] Otto, A. 1968. Excitation of nonradiative surface plasma waves in silver by the method of frustrated total reflection, *Z. Phys.*, 216:398–410. DOI: 10.1007/bf01391532. 4

[54] Pulsifer, D. P., Faryad, M., and Lakhtakia, A. 2012. Grating-coupled excitation of Tamm waves, *J. Opt. Soc. Am. B*, 29:2260–2269. DOI: 10.1364/josab.29.002260. 5

[55] Faryad, M. and Lakhtakia, A. 2011. Multiple trains of same-color surface plasmon-polaritons guided by the planar interface of a metal and a sculptured nematic thin film. Part V: Grating-coupled excitation, *J. Nanophoton.*, 5:053527. DOI: 10.1117/1.3663210. 5

[56] Hasan, M. Z. and Kane, C. L. 2010. Topological insulators, *Rev. Mod. Phys.*, 82:3045–3067. DOI: 10.1103/RevModPhys.82.3045. 5

[57] Lakhtakia, A. and Mackay, T. G. 2016. Classical electromagnetic model of surface states in topological insulators, *J. Nanophoton.*, 10:033004. DOI: 10.1117/1.jnp.10.033004. 5

[58] Diovisalvi, A., Lakhtakia, A., Fiumara, V., and Chiadini, F. 2017. Bilaterally asymmetric reflection and transmission of light by a grating structure containing a topological insulator, *Opt. Commun.*, 398:67–76. DOI: 10.1016/j.optcom.2017.04.017. 5

[59] Depine, R. A. 2016. *Graphene Optics: Electromagnetic Solution of Canonical Problems* (San Rafael, CA, Morgan & Claypool), IOP Concise Physics. DOI: 10.1088/978-1-6817-4309-7. 5

[60] Chiadini, F., Scaglione, A., Fiumara, V., Shuba, M. V., and Lakhtakia, A. 2019. Effect of chemical potential on Dyakonov–Tamm waves guided by a graphene-coated structurally chiral medium, *J. Opt.*, 21:055002, UK. DOI: 10.1088/2040-8986/ab137f.
Chiadini, F., Scaglione, A., Fiumara, V., Shuba, M. V., and Lakhtakia, A. 2019. Effect of chemical potential on Dyakonov–Tamm waves guided by a graphene-coated structurally chiral medium, *J. Opt.*, 21:079501, UK. (erratum) DOI: 10.1088/2040-8986/ab2a57. 5

[61] Chen, H. C. 1983. *Theory of Electromagnetic Waves* (New York, McGraw-Hill). 5

CHAPTER 2

Electromagnetic Preliminaries

As a precursor to the presentation of the TMM for electromagnetic reflection–transmission boundary-value problems, the essential background electromagnetic theory is presented in this chapter. After the introduction of the Maxwell postulates, the constitutive relations are described for the most general linear materials and specializations thereof. The matter of planewave propagation in such materials is then considered, leading to the formulation of a 4×4 matrix ordinary differential equation which is the bedrock of the TMM.

2.1 MAXWELL POSTULATES

2.1.1 MICROSCOPIC PERSPECTIVE

Every material is spatially nonhomogeneous and temporally varying from the *microscopic* electromagnetic perspective, since it is a collection of point charges. Given that the approach adopted in this book is a classical one, uncertainties in the positions or velocities of these point charges are irrelevant. An ensemble of point charges q_ℓ, $\ell \in \{1, 2, 3, \ldots\}$, positioned at $\underline{r}_\ell(t)$ and moving with velocity $\underline{v}_\ell(t)$ at time t, gives rise to the microscopic charge density

$$\tilde{c}(\underline{r}, t) = \sum_\ell q_\ell \, \delta\left[\underline{r} - \underline{r}_\ell(t)\right] \tag{2.1}$$

and the microscopic current density

$$\underline{\tilde{j}}(\underline{r}, t) = \sum_\ell q_\ell \, \underline{v}_\ell \, \delta\left[\underline{r} - \underline{r}_\ell(t)\right], \tag{2.2}$$

wherein the Dirac delta satisfies the constraint

$$\int_{-\infty}^{\infty} \delta(s) \, ds = 1. \tag{2.3}$$

Both of the densities are sources of two microscopic electromagnetic fields, namely the electric field $\underline{\tilde{e}}(\underline{r}, t)$ and the magnetic field $\underline{\tilde{b}}(\underline{r}, t)$.

The relationships between the microscopic source densities, $\tilde{c}(\underline{r}, t)$ and $\underline{\tilde{j}}(\underline{r}, t)$, and the fields, $\underline{\tilde{e}}(\underline{r}, t)$ and $\underline{\tilde{b}}(\underline{r}, t)$, are encapsulated by the microscopic Maxwell postulates [1]

$$\left.\begin{aligned}
\nabla \times \underline{\tilde{e}}(\underline{r}, t) + \frac{\partial}{\partial t} \underline{\tilde{b}}(\underline{r}, t) &= \underline{0} \\
\nabla \times \underline{\tilde{b}}(\underline{r}, t) - \varepsilon_0 \mu_0 \frac{\partial}{\partial t} \underline{\tilde{e}}(\underline{r}, t) &= \mu_0 \underline{\tilde{j}}(\underline{r}, t) \\
\underline{\nabla} \cdot \underline{\tilde{e}}(\underline{r}, t) &= \frac{1}{\varepsilon_0} \tilde{c}(\underline{r}, t) \\
\underline{\nabla} \cdot \underline{\tilde{b}}(\underline{r}, t) &= 0
\end{aligned}\right\}. \tag{2.4}$$

The permittivity and permeability of matter-free space are denoted by $\varepsilon_0 = 8.854 \times 10^{-12}$ F m^{-1} and $\mu_0 = 4\pi \times 10^{-7}$ H m^{-1}, respectively. The microscopic fields $\underline{\tilde{e}}(\underline{r}, t)$ and $\underline{\tilde{b}}(\underline{r}, t)$ possess spatial variations over distances $\lesssim 10^{-10}$ m and temporal variations over durations ranging from $\lesssim 10^{-13}$ s for nuclear vibrations to $\lesssim 10^{-17}$ s for electronic orbital motion [1].

2.1.2   MACROSCOPIC PERSPECTIVE

The summation index ℓ in (2.1) and (2.2) ranges from 1 to an impractically large number in any volume occupied by a material that may be characterized from a macroscopic perspective. Consequently, it is desirable to consider the spatiotemporal averages of the microscopic quantities in (2.4), from a *macroscopic* perspective [2]. In fact, spatial averaging alone suffices since it implicitly involves temporal averaging, due to the universal maximum speed $c_0 = (\varepsilon_0 \mu_0)^{-1/2}$ being finite [1]. The spatial average should be taken over volumes which are sufficiently large as to contain many point charges but the linear extent of the averaging volume should be much smaller than the smallest electromagnetic wavelength in consideration. Thus, the macroscopic perspective is appropriate for solids and liquids in the extreme-ultraviolet regime and in longer-wavelength regimes. The appropriate wavelength range for the macroscopic perspective may have a much larger lower bound in rarefied gases.

Adoption of the macroscopic perspective leads to the replacement of (2.4) by

$$\left.\begin{aligned}
\nabla \times \underline{\tilde{E}}(\underline{r}, t) + \frac{\partial}{\partial t} \underline{\tilde{B}}(\underline{r}, t) &= \underline{0} \\
\nabla \times \underline{\tilde{B}}(\underline{r}, t) - \varepsilon_0 \mu_0 \frac{\partial}{\partial t} \underline{\tilde{E}}(\underline{r}, t) &= \mu_0 \underline{\tilde{J}}(\underline{r}, t) \\
\underline{\nabla} \cdot \underline{\tilde{E}}(\underline{r}, t) &= \frac{1}{\varepsilon_0} \tilde{\rho}(\underline{r}, t) \\
\underline{\nabla} \cdot \underline{\tilde{B}}(\underline{r}, t) &= 0
\end{aligned}\right\}. \tag{2.5}$$

The macroscopic fields $\underline{\tilde{E}}(\underline{r}, t)$ and $\underline{\tilde{B}}(\underline{r}, t)$ represent the spatial averages of $\underline{\tilde{e}}(\underline{r}, t)$ and $\underline{\tilde{b}}(\underline{r}, t)$, respectively; and the macroscopic charge and current densities $\tilde{\rho}(\underline{r}, t)$ and $\underline{\tilde{J}}(\underline{r}, t)$ are likewise related to $\tilde{c}(\underline{r}, t)$ and $\underline{\tilde{j}}(\underline{r}, t)$.

The macroscopic source densities in a material may be regarded as either externally impressed or arising from internal mechanisms. Thus, these source densities can be partitioned as

$$\left. \begin{aligned} \tilde{\rho}(\underline{r}, t) &= \tilde{\rho}_{\text{ext}}(\underline{r}, t) + \tilde{\rho}_{\text{int}}(\underline{r}, t) \\ \underline{\tilde{J}}(\underline{r}, t) &= \underline{\tilde{J}}_{\text{ext}}(\underline{r}, t) + \underline{\tilde{J}}_{\text{int}}(\underline{r}, t) \end{aligned} \right\}, \tag{2.6}$$

wherein the subscripts "ext" and "int" identify the externally impressed and internally arising source densities, respectively. The internally arising source densities are characterized via the macroscopic polarization $\underline{\tilde{P}}(\underline{r}, t)$ and magnetization $\underline{\tilde{M}}(\underline{r}, t)$ as follows:

$$\left. \begin{aligned} \tilde{\rho}_{\text{int}}(\underline{r}, t) &= -\underline{\nabla} \cdot \underline{\tilde{P}}(\underline{r}, t) \\ \underline{\tilde{J}}_{\text{int}}(\underline{r}, t) &= \frac{\partial}{\partial t} \underline{\tilde{P}}(\underline{r}, t) + \frac{1}{\mu_0} \underline{\nabla} \times \underline{\tilde{M}}(\underline{r}, t) \end{aligned} \right\}. \tag{2.7}$$

However, if $\underline{\tilde{P}}(\underline{r}, t)$ were replaced by $\underline{\tilde{P}}(\underline{r}, t) - \underline{\nabla} \times \underline{\tilde{A}}(\underline{r}, t)$ and $\underline{\tilde{M}}(\underline{r}, t)$ by $\underline{\tilde{M}}(\underline{r}, t) + \mu_0 (\partial/\partial t) \underline{\tilde{A}}(\underline{r}, t)$, where $\underline{\tilde{A}}(\underline{r}, t)$ is some differentiable vector function, then $\tilde{\rho}_{\text{int}}(\underline{r}, t)$ and $\underline{\tilde{J}}_{\text{int}}(\underline{r}, t)$ given by (2.7) would remain unchanged. Hence, a degree of ambiguity is associated with the internally arising source densities represented by (2.7).

The polarization and magnetization are subsumed into the following definitions of two macroscopic electromagnetic fields:[1]

$$\left. \begin{aligned} \underline{\tilde{D}}(\underline{r}, t) &= \varepsilon_0 \underline{\tilde{E}}(\underline{r}, t) + \underline{\tilde{P}}(\underline{r}, t) \\ \underline{\tilde{H}}(\underline{r}, t) &= \frac{1}{\mu_0} \underline{\tilde{B}}(\underline{r}, t) - \frac{1}{\mu_0} \underline{\tilde{M}}(\underline{r}, t) \end{aligned} \right\}. \tag{2.8}$$

The fields $\underline{\tilde{D}}(\underline{r}, t)$ and $\underline{\tilde{H}}(\underline{r}, t)$ develop in a material in response to the fields $\underline{\tilde{E}}(\underline{r}, t)$ and $\underline{\tilde{B}}(\underline{r}, t)$. Consequently, $\underline{\tilde{D}}(\underline{r}, t)$ and $\underline{\tilde{H}}(\underline{r}, t)$ are regarded as *induction* fields, while $\underline{\tilde{E}}(\underline{r}, t)$ and $\underline{\tilde{B}}(\underline{r}, t)$ are regarded as *primitive* fields. Unlike the induction fields, the primitive fields may be measured directly via the Lorentz force [1]

$$\underline{\tilde{F}}_{\text{Lor}}(\underline{r}, t) = q(\underline{r}, t) \left[\underline{\tilde{E}}(\underline{r}, t) + \underline{v}(\underline{r}, t) \times \underline{\tilde{B}}(\underline{r}, t) \right] \tag{2.9}$$

acting on a point charge $q(\underline{r}, t)$ traveling at velocity $\underline{v}(\underline{r}, t)$.

By substituting (2.6)–(2.8) into (2.5), the four macroscopic electromagnetic fields $\underline{\tilde{E}}(\underline{r}, t)$, $\underline{\tilde{D}}(\underline{r}, t)$, $\underline{\tilde{B}}(\underline{r}, t)$, and $\underline{\tilde{H}}(\underline{r}, t)$ are brought together in the macroscopic Maxwell postulates written

---

[1]An alternative convention is in common use concerning the definitions of the vectors $\underline{\tilde{J}}_{\text{int}}(\underline{r}, t)$, $\underline{\tilde{M}}(\underline{r}, t)$, and $\underline{\tilde{H}}(\underline{r}, t)$. This convention involves (2.7)$_2$ being replaced by $\underline{\tilde{J}}_{\text{int}}(\underline{r}, t) = (\partial/\partial t) \underline{\tilde{P}}(\underline{r}, t) + \underline{\nabla} \times \underline{\tilde{M}}(\underline{r}, t)$ and (2.8)$_2$ being replaced by $\underline{\tilde{H}}(\underline{r}, t) = \mu_0^{-1} \underline{\tilde{B}}(\underline{r}, t) - \underline{\tilde{M}}(\underline{r}, t)$. The alternative convention is not used in this book.

in standard form as follows:

$$\left.\begin{array}{c} \nabla \times \underline{\tilde{H}}(\underline{r},t) - \dfrac{\partial}{\partial t}\,\underline{\tilde{D}}(\underline{r},t) = \underline{\tilde{J}}_{ext}(\underline{r},t) \\[2mm] \nabla \times \underline{\tilde{E}}(\underline{r},t) + \dfrac{\partial}{\partial t}\,\underline{\tilde{B}}(\underline{r},t) = \underline{0} \\[2mm] \underline{\nabla} \cdot \underline{\tilde{D}}(\underline{r},t) = \tilde{\rho}_{ext}(\underline{r},t) \\[2mm] \underline{\nabla} \cdot \underline{\tilde{B}}(\underline{r},t) = 0 \end{array}\right\}. \tag{2.10}$$

These postulates comprise four linear differential equations: (i) the two homogeneous differential equations $(2.10)_2$ and $(2.10)_4$ involving the primitive fields and (ii) the two inhomogeneous differential equations $(2.10)_1$ and $(2.10)_3$ involving the induction fields.

The divergence postulate $(2.10)_4$ is completely consistent with the curl postulate $(2.10)_2$, as can be seen by taking the divergence of the left and right sides of the latter postulate. The consistency of the divergence postulate $(2.10)_3$ and the curl postulate $(2.10)_1$ mandates the reasonable constraint

$$\underline{\nabla} \cdot \underline{\tilde{J}}_{ext}(\underline{r},t) + \frac{\partial}{\partial t}\,\tilde{\rho}_{ext}(\underline{r},t) = 0. \tag{2.11}$$

The two Maxwell curl postulates suffice for our purposes in the remainder of this book, the macroscopic continuity equation (2.11) being presumed to hold in practical situations.

2.2   CONSTITUTIVE RELATIONS

Although the Maxwell postulates (2.10) govern all electromagnetic phenomenons in materials, they cannot be solved (for the primitive or the induction fields) without further information being supplied. This further information is provided by constitutive relations, which relate the primitive fields to the induction fields in the material under consideration. Formally, these may be expressed in the general form

$$\left.\begin{array}{c} \underline{\tilde{D}}(\underline{r},t) = \mathcal{F}\{\underline{\tilde{E}}(\underline{r},t), \underline{\tilde{B}}(\underline{r},t)\} \\[2mm] \underline{\tilde{H}}(\underline{r},t) = \mathcal{G}\{\underline{\tilde{E}}(\underline{r},t), \underline{\tilde{B}}(\underline{r},t)\} \end{array}\right\}, \tag{2.12}$$

with $\mathcal{F}$ and $\mathcal{G}$ being linear functions of $\underline{\tilde{E}}(\underline{r},t)$ and $\underline{\tilde{B}}(\underline{r},t)$ for linear materials, and nonlinear functions of $\underline{\tilde{E}}(\underline{r},t)$ and $\underline{\tilde{B}}(\underline{r},t)$ for nonlinear materials. This book is devoted to the TMM for linear materials.

In general, a material's electromagnetic response is spatiotemporally nonlocal. Consequently, in full generality, the constitutive relations of a linear material are expressed as [3]

$$
\left.
\begin{aligned}
\tilde{\underline{D}}(\underline{r},t) &= \int_{t'}\int_{\underline{r}'}\left[\tilde{\underline{\underline{\varepsilon}}}_{\mathrm{EB}}(\underline{r}',t')\cdot\tilde{\underline{E}}(\underline{r}-\underline{r}',t-t')\right.\\
&\quad \left.+\tilde{\underline{\underline{\xi}}}_{\mathrm{EB}}(\underline{r}',t')\cdot\tilde{\underline{B}}(\underline{r}-\underline{r}',t-t')\right]d^3\underline{r}'\,dt'\\
\tilde{\underline{H}}(\underline{r},t) &= \int_{t'}\int_{\underline{r}'}\left[\tilde{\underline{\underline{\zeta}}}_{\mathrm{EB}}(\underline{r}',t')\cdot\tilde{\underline{E}}(\underline{r}-\underline{r}',t-t')\right.\\
&\quad \left.+\tilde{\underline{\underline{\nu}}}_{\mathrm{EB}}(\underline{r}',t')\cdot\tilde{\underline{B}}(\underline{r}-\underline{r}',t-t')\right]d^3\underline{r}'\,dt'
\end{aligned}
\right\},
\qquad (2.13)
$$

in terms of the four 3×3 constitutive dyadics $\tilde{\underline{\underline{\varepsilon}}}_{\mathrm{EB}}(\underline{r},t), \tilde{\underline{\underline{\xi}}}_{\mathrm{EB}}(\underline{r},t), \tilde{\underline{\underline{\zeta}}}_{\mathrm{EB}}(\underline{r},t)$, and $\tilde{\underline{\underline{\nu}}}_{\mathrm{EB}}(\underline{r},t)$. A guide to 3×3 dyadics is provided in Appendix A.

The manifestation of nonlocality can vary greatly from one material to another. If a characteristic length scale in a material is similar to the electromagnetic wavelength, then spatial nonlocality may be significant [4]. However, the effects of spatial nonlocality are negligible in the vast majority of situations currently considered as practical. On the other hand, since electromagnetic signals travel rapidly, the effects of temporal nonlocality must not be ignored. Spatially local but temporally nonlocal linear materials are characterized by the following constitutive relations:

$$
\left.
\begin{aligned}
\tilde{\underline{D}}(\underline{r},t) &= \int_{t'}\left[\tilde{\underline{\underline{\varepsilon}}}_{\mathrm{EB}}(\underline{r},t')\cdot\tilde{\underline{E}}(\underline{r},t-t')+\tilde{\underline{\underline{\xi}}}_{\mathrm{EB}}(\underline{r},t')\cdot\tilde{\underline{B}}(\underline{r},t-t')\right]dt'\\
\tilde{\underline{H}}(\underline{r},t) &= \int_{t'}\left[\tilde{\underline{\underline{\zeta}}}_{\mathrm{EB}}(\underline{r},t')\cdot\tilde{\underline{E}}(\underline{r},t-t')+\tilde{\underline{\underline{\nu}}}_{\mathrm{EB}}(\underline{r},t')\cdot\tilde{\underline{B}}(\underline{r},t-t')\right]dt'
\end{aligned}
\right\}.
\qquad (2.14)
$$

2.3   FREQUENCY DOMAIN

The time-domain constitutive relations (2.14) are often inconvenient as the convolution integrals therein usually lead to mathematical complications that are analytically intractable and computationally challenging. These complications are bypassed by switching to the frequency domain by taking the temporal Fourier transforms of all fields and sources appearing in (2.10) and (2.14) as follows:

$$
\mathcal{Z}(\underline{r},\omega) = \int_{-\infty}^{\infty}\tilde{\mathcal{Z}}(\underline{r},t)\exp(i\omega t)\,dt, \quad \mathcal{Z}\in\left\{\underline{\underline{\varepsilon}}_{\mathrm{EB}},\underline{\underline{\xi}}_{\mathrm{EB}},\underline{\underline{\zeta}}_{\mathrm{EB}},\underline{\underline{\nu}}_{\mathrm{EB}},\underline{E},\underline{D},\underline{B},\underline{H},\underline{J}_{\mathrm{ext}},\rho_{\mathrm{ext}}\right\}.
\quad (2.15)
$$

Here, ω is the angular frequency and $i=\sqrt{-1}$. When $\tilde{\mathcal{Z}}(\underline{r},t)$ is either a source or a field, its counterpart $\mathcal{Z}(\underline{r},\omega)$ is referred to as a *phasor*.

Accordingly, the Maxwell postulates (2.10) transform in the frequency domain to

$$
\left.
\begin{aligned}
\nabla \times \underline{H}(\underline{r},\omega) + i\omega \underline{D}(\underline{r},\omega) &= \underline{J}_{\text{ext}}(\underline{r},\omega) \\
\nabla \times \underline{E}(\underline{r},\omega) - i\omega \underline{B}(\underline{r},\omega) &= \underline{0} \\
\underline{\nabla} \cdot \underline{D}(\underline{r},\omega) &= \rho_{\text{ext}}(\underline{r},\omega) \\
\underline{\nabla} \cdot \underline{B}(\underline{r},\omega) &= 0
\end{aligned}
\right\} ; \tag{2.16}
$$

furthermore, application of the convolution theorem [5] to (2.14) delivers the frequency-domain constitutive relations

$$
\left.
\begin{aligned}
\underline{D}(\underline{r},\omega) &= \underline{\underline{\varepsilon}}_{\text{EB}}(\underline{r},\omega) \cdot \underline{E}(\underline{r},\omega) + \underline{\underline{\xi}}_{\text{EB}}(\underline{r},\omega) \cdot \underline{B}(\underline{r},\omega) \\
\underline{H}(\underline{r},\omega) &= \underline{\underline{\zeta}}_{\text{EB}}(\underline{r},\omega) \cdot \underline{E}(\underline{r},\omega) + \underline{\underline{\nu}}_{\text{EB}}(\underline{r},\omega) \cdot \underline{B}(\underline{r},\omega)
\end{aligned}
\right\} . \tag{2.17}
$$

A price has to be paid for the relative simplicity of the frequency-domain constitutive relations (2.17) as compared with the time-domain constitutive relations (2.14): The electromagnetic fields and constitutive dyadics in the time-domain description (2.14) are all real-valued quantities whereas their counterparts in the frequency-domain description (2.17) are all complex valued. Thus, inverse temporal Fourier transforms must be taken to convert the frequency-domain quantities into real-valued physical quantities.

The principle of causality leads to constraints on the real and imaginary parts of the frequency-domain constitutive parameters. Since a passive material cannot respond to a stimulus until *after* the stimulus has been received, we have

$$
\left.
\begin{aligned}
\underline{\underline{\tilde{\varepsilon}}}_{\text{EB}}(\underline{r},t) - \varepsilon_0 \delta(t)\underline{\underline{I}} &\equiv \underline{\underline{0}} \\
\underline{\underline{\tilde{\xi}}}_{\text{EB}}(\underline{r},t) &\equiv \underline{\underline{0}} \\
\underline{\underline{\tilde{\zeta}}}_{\text{EB}}(\underline{r},t) &\equiv \underline{\underline{0}} \\
\mu_0^{-1}\delta(t)\underline{\underline{I}} - \underline{\underline{\tilde{\nu}}}_{\text{EB}}(\underline{r},t) &\equiv \underline{\underline{0}}
\end{aligned}
\right\} \quad \text{for} \quad t \leq 0. \tag{2.18}
$$

Now let the scalar function $\tilde{\mathcal{Z}}(\underline{r},t)$ represent an arbitrary component of any one of the dyadics $\underline{\underline{\tilde{\varepsilon}}}_{\text{EB}}(\underline{r},t) - \varepsilon_0\delta(t)\underline{\underline{I}}$, $\underline{\underline{\tilde{\xi}}}_{\text{EB}}(\underline{r},t)$, $\underline{\underline{\tilde{\zeta}}}_{\text{EB}}(\underline{r},t)$, and $\mu_0^{-1}\delta(t)\underline{\underline{I}} - \underline{\underline{\tilde{\nu}}}_{\text{EB}}(\underline{r},t)$. The analytic continuation of $\mathcal{Z}(\underline{r},\omega)$ in the upper half of the complex-ω plane then yields the Hilbert transforms

$$
\left.
\begin{aligned}
\text{Re}\,\{\mathcal{Z}(\underline{r},\omega)\} &= \frac{1}{\pi}\text{P}\int_{-\infty}^{\infty} \frac{\text{Im}\,\{\mathcal{Z}(\underline{r},s)\}}{s-\omega}\,ds \\
\text{Im}\,\{\mathcal{Z}(\underline{r},\omega)\} &= -\frac{1}{\pi}\text{P}\int_{-\infty}^{\infty} \frac{\text{Re}\,\{\mathcal{Z}(\underline{r},s)\}}{s-\omega}\,ds
\end{aligned}
\right\} , \tag{2.19}
$$

where P signifies the Cauchy principal value. As $\tilde{\mathcal{Z}}(\underline{r}, t)$ is real valued, $\mathcal{Z}(\underline{r}, \omega)$ is related to its complex conjugate $\mathcal{Z}^*(\underline{r}, \omega)$ by

$$\mathcal{Z}(\underline{r}, -\omega) = \mathcal{Z}^*(\underline{r}, \omega). \tag{2.20}$$

By exploiting the symmetry condition (2.20), the Hilbert transforms (2.19) give rise to the Kramers–Kronig relations [6]

$$\left. \begin{aligned} \text{Re}\,\{\mathcal{Z}(\underline{r}, \omega)\} &= \frac{2}{\pi} P \int_0^\infty \frac{s\,\text{Im}\,\{\mathcal{Z}(\underline{r}, s)\}}{s^2 - \omega^2}\,ds \\ \text{Im}\,\{\mathcal{Z}(\underline{r}, \omega)\} &= -\frac{2}{\pi} P \int_0^\infty \frac{\omega\,\text{Re}\,\{\mathcal{Z}(\underline{r}, s)\}}{s^2 - \omega^2}\,ds \end{aligned} \right\}. \tag{2.21}$$

These two relations are representative of general constraints on the frequency responses of causal linear systems [7].

The partnering of the primitive field phasors $\underline{E}(\underline{r}, \omega)$ and $\underline{B}(\underline{r}, \omega)$ leads to the Boys–Post formulation of the constitutive relations equations (2.17). The field phasors $\underline{E}(\underline{r}, \omega)$ and $\underline{B}(\underline{r}, \omega)$ are paired because their inverse temporal Fourier transforms appear in the Lorentz force. Alternatively, it can be reasonable to partner $\underline{E}(\underline{r}, \omega)$ with $\underline{H}(\underline{r}, \omega)$, in light of the standard boundary conditions as well as the definition of the time-averaged Poynting vector [8]. That pairing leads to the Tellegen formulation of the frequency-domain constitutive relations

$$\left. \begin{aligned} \underline{D}(\underline{r}, \omega) &= \underline{\underline{\varepsilon}}_{EH}(\underline{r}, \omega) \cdot \underline{E}(\underline{r}, \omega) + \underline{\underline{\xi}}_{EH}(\underline{r}, \omega) \cdot \underline{H}(\underline{r}, \omega) \\ \underline{B}(\underline{r}, \omega) &= \underline{\underline{\zeta}}_{EH}(\underline{r}, \omega) \cdot \underline{E}(\underline{r}, \omega) + \underline{\underline{\mu}}_{EH}(\underline{r}, \omega) \cdot \underline{H}(\underline{r}, \omega) \end{aligned} \right\}. \tag{2.22}$$

The constitutive dyadics in the Boys–Post equations (2.17) are related to those in the Tellegen equations (2.22) as follows [3]:

$$\left. \begin{aligned} \underline{\underline{\varepsilon}}_{EH}(\underline{r}, \omega) &= \underline{\underline{\varepsilon}}_{EB}(\underline{r}, \omega) - \underline{\underline{\xi}}_{EB}(\underline{r}, \omega) \cdot \underline{\underline{\nu}}_{EB}^{-1}(\underline{r}, \omega) \cdot \underline{\underline{\zeta}}_{EB}(\underline{r}, \omega) \\ \underline{\underline{\xi}}_{EH}(\underline{r}, \omega) &= \underline{\underline{\xi}}_{EB}(\underline{r}, \omega) \cdot \underline{\underline{\nu}}_{EB}^{-1}(\underline{r}, \omega) \\ \underline{\underline{\zeta}}_{EH}(\underline{r}, \omega) &= -\underline{\underline{\nu}}_{EB}^{-1}(\underline{r}, \omega) \cdot \underline{\underline{\zeta}}_{EB}(\underline{r}, \omega) \\ \underline{\underline{\mu}}_{EH}(\underline{r}, \omega) &= \underline{\underline{\nu}}_{EB}^{-1}(\underline{r}, \omega) \end{aligned} \right\}. \tag{2.23}$$

Implicitly, the constitutive dyadic $\underline{\underline{\nu}}_{EB}(\underline{r}, \omega)$ is nonsingular. Caution should be exercised here, since singular constitutive dyadics are known to occur [9], albeit rarely.

The Tellegen constitutive relations (2.22) are used in the remainder of this book. For brevity, the subscript "EH" on the constitutive dyadics is omitted. For the same reason, ω is omitted from the argument list for the electromagnetic fields and sources. Thus, the Tellegen constitutive relations are written as

$$\left. \begin{aligned} \underline{D}(\underline{r}) &= \underline{\underline{\varepsilon}}(\underline{r}) \cdot \underline{E}(\underline{r}) + \underline{\underline{\xi}}(\underline{r}) \cdot \underline{H}(\underline{r}) \\ \underline{B}(\underline{r}) &= \underline{\underline{\zeta}}(\underline{r}) \cdot \underline{E}(\underline{r}) + \underline{\underline{\mu}}(\underline{r}) \cdot \underline{H}(\underline{r}) \end{aligned} \right\}. \tag{2.24}$$

If a material is spatially homogeneous, its constitutive relations are written even more simply as

$$\left. \begin{array}{l} \underline{D}(\underline{r}) = \underline{\underline{\varepsilon}} \cdot \underline{E}(\underline{r}) + \underline{\underline{\xi}} \cdot \underline{H}(\underline{r}) \\ \underline{B}(\underline{r}) = \underline{\underline{\zeta}} \cdot \underline{E}(\underline{r}) + \underline{\underline{\mu}} \cdot \underline{H}(\underline{r}) \end{array} \right\} . \tag{2.25}$$

2.4   CONSTITUTIVE DYADICS

As is clear from the Tellegen constitutive relations (2.25), the most general linear homogeneous material is characterized by four 3×3 constitutive dyadics—namely, the permittivity dyadic $\underline{\underline{\varepsilon}}$, the permeability dyadic $\underline{\underline{\mu}}$, and the two magnetoelectric dyadics $\underline{\underline{\xi}}$ and $\underline{\underline{\zeta}}$. Thus, a total of 36 complex-valued constitutive parameters specify a general linear material. Spacetime symmetries greatly reduce the dimensionality of the constitutive-parameter space for many materials of interest [10], as illustrated in the following subsections.

2.4.1   ISOTROPIC AND BI-ISOTROPIC MATERIALS

In isotropic dielectric-magnetic materials, the induction fields are aligned wholly parallel to the primitive fields. Hence, their constitutive dyadics are of the form

$$\left. \begin{array}{l} \underline{\underline{\varepsilon}} = \varepsilon \underline{\underline{I}} \\ \underline{\underline{\xi}} = \underline{\underline{0}} \\ \underline{\underline{\zeta}} = \underline{\underline{0}} \\ \underline{\underline{\mu}} = \mu \underline{\underline{I}} \end{array} \right\} , \qquad \varepsilon \in \mathbb{C}, \quad \mu \in \mathbb{C}, \tag{2.26}$$

where $\mathbb{C}$ is the set of all complex numbers. Furthermore, $\mu = \mu_0$ for isotropic dielectric materials and $\varepsilon = \varepsilon_0$ for isotropic magnetic materials.

A bi-isotropic material displays both isotropy and magnetoelectric coupling. The constitutive dyadics of bi-isotropic materials are of the form

$$\left. \begin{array}{l} \underline{\underline{\varepsilon}} = \varepsilon \underline{\underline{I}} \\ \underline{\underline{\xi}} = \xi \underline{\underline{I}} \\ \underline{\underline{\zeta}} = \zeta \underline{\underline{I}} \\ \underline{\underline{\mu}} = \mu \underline{\underline{I}} \end{array} \right\} , \qquad \varepsilon \in \mathbb{C}, \quad \xi \in \mathbb{C}, \quad \zeta \in \mathbb{C}, \quad \mu \in \mathbb{C} . \tag{2.27}$$

An important special case occurs if $\xi = -\zeta \neq 0$; then the material is an isotropic chiral material or a reciprocal bi-isotropic material [11]. The *optical activity* of such materials can be harnessed to discriminate between left-handed and right-handed electromagnetic fields [12]. The case $\xi = \zeta \neq 0$ is also a notable one, at least from a theoretical standpoint. This is the case for topological insulators [13], but it is physically appropriate to incorporate surface states leading to jump conditions across interfaces and setting $\xi = \zeta = 0$ [14].

Although the 3×3 constitutive dyadics of bi-isotropic materials are simply scalar multiples of $\underline{\underline{I}}$, the induction fields are generally not aligned with the primitive fields in these materials, unlike in isotropic dielectric-magnetic materials. However, key electromagnetic characteristics, such as those pertaining to planewave propagation, are independent of direction in bi-isotropic materials.

2.4.2   ANISOTROPIC AND BIANISOTROPIC MATERIALS

Many naturally occurring and engineered materials exhibit anisotropy, i.e., direction-dependent characteristics [15]. Nontrivial dyadics—as opposed to dyadics that are simply scalar multiples of $\underline{\underline{I}}$—are needed to relate the induction field phasors to the primitive field phasors for such materials.

In an anisotropic dielectric-magnetic material, the induction field $\underline{D}$ is anisotropically coupled to the primitive field $\underline{E}$ and/or the induction field $\underline{H}$ is anisotropically coupled to the primitive field $\underline{B}$. Hence, the constitutive dyadics obey the following constraints:

$$\left.\begin{array}{l} \underline{\underline{\varepsilon}} \neq \varepsilon\underline{\underline{I}} \\[4pt] \underline{\underline{\xi}} = \underline{\underline{0}} \\[4pt] \underline{\underline{\zeta}} = \underline{\underline{0}} \\[4pt] \underline{\underline{\mu}} \neq \mu\underline{\underline{I}} \end{array}\right\}. \tag{2.28}$$

Whereas $\underline{\underline{\mu}} = \mu_0\underline{\underline{I}}$ for an anisotropic dielectric material, $\underline{\underline{\varepsilon}} = \varepsilon_0\underline{\underline{I}}$ for an anisotropic magnetic material. For a general anisotropic dielectric-magnetic material, the non-trivial constitutive dyadics may be represented by the 3×3 Cartesian matrixes

$$\left.\begin{array}{l} \underline{\underline{\varepsilon}} = \left[\begin{array}{ccc} \varepsilon_{xx} & \varepsilon_{xy} & \varepsilon_{xz} \\ \varepsilon_{yx} & \varepsilon_{yy} & \varepsilon_{yz} \\ \varepsilon_{zx} & \varepsilon_{zy} & \varepsilon_{zz} \end{array}\right] \\[30pt] \underline{\underline{\mu}} = \left[\begin{array}{ccc} \mu_{xx} & \mu_{xy} & \mu_{xz} \\ \mu_{yx} & \mu_{yy} & \mu_{yz} \\ \mu_{zx} & \mu_{zy} & \mu_{zz} \end{array}\right] \end{array}\right\}, \tag{2.29}$$

with all matrix entries being complex valued.

The natural generalization of anisotropy is bianisotropy. In a bianisotropic material, both induction fields $\underline{D}$ and $\underline{H}$ are anisotropically coupled to both primitive fields $\underline{E}$ and $\underline{B}$. Thus, the constitutive dyadics for a bianisotropic material satisfy

$$\left.\begin{array}{r} \underline{\underline{\varepsilon}} \neq \varepsilon\underline{\underline{I}} \\[4pt] \underline{\underline{\xi}} \neq \xi\underline{\underline{I}} \\[4pt] \underline{\underline{\zeta}} \neq \zeta\underline{\underline{I}} \\[4pt] \underline{\underline{\mu}} \neq \mu\underline{\underline{I}} \end{array}\right\}. \tag{2.30}$$

For a general bianisotropic material, the constitutive dyadics may be represented by the 3×3 Cartesian matrixes (2.29) together with

$$\left.\begin{array}{l} \underline{\underline{\xi}} = \left[\begin{array}{ccc} \xi_{xx} & \xi_{xy} & \xi_{xz} \\ \xi_{yx} & \xi_{yy} & \xi_{yz} \\ \xi_{zx} & \xi_{zy} & \xi_{zz} \end{array}\right] \\[24pt] \underline{\underline{\zeta}} = \left[\begin{array}{ccc} \zeta_{xx} & \zeta_{xy} & \zeta_{xz} \\ \zeta_{yx} & \zeta_{yy} & \zeta_{yz} \\ \zeta_{zx} & \zeta_{zy} & \zeta_{zz} \end{array}\right] \end{array}\right\}, \tag{2.31}$$

with all matrix entries being complex valued.

The structures of the constitutive dyadics in (2.28) and (2.30) reflect underlying spacetime symmetries of the materials being characterized. These symmetries are conventionally classified in terms of 122 discrete magnetic point groups and 21 continuous magnetic point groups (which encompass the symmetries of isotropic and bi-isotropic materials as degenerate cases) [10, 16–18]. The magnetic point group with the lowest symmetry, namely C_1 in the Schoenflies notation, is especially noteworthy. For materials in this magnetic point group, the structures of the four constitutive dyadics $\underline{\underline{\varepsilon}}$, $\underline{\underline{\xi}}$, $\underline{\underline{\zeta}}$, and $\underline{\underline{\mu}}$ are completely arbitrary; i.e., a total of 36 complex-valued constitutive scalars are necessary. However, most commonly encountered anisotropic and bianisotropic materials exhibit substantial degrees of spacetime symmetry and, accordingly, far fewer constitutive scalars are needed to characterize them.

The simplest form of anisotropy is exemplified by uniaxial materials, for which the constitutive dyadics take the form

$$\underline{\underline{\sigma}} = \sigma_1\underline{\underline{I}} + \sigma_2\underline{u}_{\mathrm{m}}\,\underline{u}_{\mathrm{m}}, \qquad \sigma_1 \in \mathbb{C}, \quad \sigma_2 \in \mathbb{C}, \quad \sigma \in \{\varepsilon, \xi, \zeta, \mu\}\,. \tag{2.32}$$

There is just one distinguished direction, identified by the unit vector $\underline{u}_{\mathrm{m}}$. In *crystal optics*, a uniaxial dielectric material is characterized by

$$\left.\begin{array}{l} \underline{\underline{\varepsilon}} = \varepsilon_1\underline{\underline{I}} + \varepsilon_2\underline{u}_{\mathrm{m}}\,\underline{u}_{\mathrm{m}} \\[6pt] \underline{\underline{\xi}} = \underline{\underline{0}} \\[6pt] \underline{\underline{\zeta}} = \underline{\underline{0}} \\[6pt] \underline{\underline{\mu}} = \mu_0\underline{\underline{I}} \end{array}\right\}, \qquad \varepsilon_1 \in \mathbb{R}, \quad \varepsilon_2 \in \mathbb{R}, \tag{2.33}$$

with $\underline{u}_{\mathrm{m}}$ pointing in the direction of the *optic axis* [8] and $\mathbb{R}$ being the set of all real numbers. That is, $\pm\underline{u}_{\mathrm{m}}$ identify the special directions along which plane waves propagate with only one phase speed. Parenthetically, $\underline{u}_{\mathrm{m}}$ is also aligned with the *optic ray axis* that represents the special direction along which plane waves propagate with only one energy speed [19].

On generalizing the anisotropy represented in (2.32), constitutive dyadics of the form

$$\underline{\underline{\sigma}} \;=\; \sigma_1\underline{\underline{I}} + \sigma_2\underline{u}_{\mathrm{m}}\,\underline{u}_{\mathrm{m}} + \sigma_3\underline{u}_{\mathrm{n}}\,\underline{u}_{\mathrm{n}},$$
$$\sigma_1 \in \mathbb{C}, \quad \sigma_2 \in \mathbb{C}, \quad \sigma_3 \in \mathbb{C}, \quad \sigma \in \{\varepsilon, \xi, \zeta, \mu\}\,, \tag{2.34}$$

emerge. These constitutive dyadics characterize orthorhombic biaxial materials. Herein the unit vectors $\underline{u}_{\mathrm{m}}$ and $\underline{u}_{\mathrm{n}}$ are mutually orthogonal. There are two distinguished directions here but the formulation (2.34) is not particularly insightful, as a physical interpretation of $\underline{u}_{\mathrm{m}}$ and $\underline{u}_{\mathrm{n}}$ is not readily forthcoming. The distinguished directions may be better appreciated by considering the following alternative formulation: for example, an orthorhombic dielectric material that is neither dissipative nor active may be characterized by [8, 20]

$$\left.\begin{aligned}
\underline{\underline{\varepsilon}} &= \varepsilon_{\mathrm{p}}\underline{\underline{I}} + \varepsilon_{\mathrm{q}}\left(\underline{u}_{\mathrm{p}}\,\underline{u}_{\mathrm{q}} + \underline{u}_{\mathrm{q}}\,\underline{u}_{\mathrm{p}}\right) \\
\underline{\underline{\xi}} &= \underline{\underline{0}} \\
\underline{\underline{\zeta}} &= \underline{\underline{0}} \\
\underline{\underline{\mu}} &= \mu_0\underline{\underline{I}}
\end{aligned}\right\}, \qquad \varepsilon_{\mathrm{p}} \in \mathbb{R}, \quad \varepsilon_{\mathrm{q}} \in \mathbb{R}. \tag{2.35}$$

The unit vectors $\underline{u}_{\mathrm{p}}$ and $\underline{u}_{\mathrm{q}}$ herein are aligned with the optic ray axes [21]; that is, they are aligned with the two directions along which plane waves propagate with only one energy speed. Notice that in the case of biaxial dielectric materials the two optic ray axes do not generally coincide with the two optic axes (which represent the directions along which plane waves propagate with only one phase speed) [10, 19].

The uniaxial and biaxial constitutive dyadics present in (2.32) and (2.34) are symmetric. Antisymmetric constitutive dyadics are also commonly encountered, notably in the context of gyrotropic materials which are characterized by constitutive dyadics of the form

$$\underline{\underline{\sigma}} \;=\; \sigma_1\underline{\underline{I}} + \sigma_2\underline{u}_{\mathrm{m}}\,\underline{u}_{\mathrm{m}} + \sigma_3\underline{u}_{\mathrm{m}} \times \underline{\underline{I}},$$
$$\sigma_1 \in \mathbb{C}, \quad \sigma_2 \in \mathbb{C}, \quad \sigma_3 \in \mathbb{C}, \quad \sigma \in \{\varepsilon, \xi, \zeta, \mu\}\,. \tag{2.36}$$

For example, a nondissipative magnetically biased ferrite may characterized by [22]

$$\left.\begin{aligned}
\underline{\underline{\varepsilon}} &= \varepsilon_0\underline{\underline{I}} \\
\underline{\underline{\xi}} &= \underline{\underline{0}} \\
\underline{\underline{\zeta}} &= \underline{\underline{0}} \\
\underline{\underline{\mu}} &= \mu_1\underline{\underline{I}} + \mu_2\underline{u}_{\mathrm{m}}\,\underline{u}_{\mathrm{m}} + i\mu_3\underline{u}_{\mathrm{m}} \times \underline{\underline{I}}
\end{aligned}\right\}, \qquad \mu_1 \in \mathbb{R}, \quad \mu_2 \in \mathbb{R}, \quad \mu_2 \in \mathbb{R}, \tag{2.37}$$

where the unit vector $\underline{u}_m$ is aligned with the direction of the biasing magnetic field. An example of gyrotropy in a bianisotropic setting arises in the case of Faraday chiral materials. These are characterized by constitutive dyadics of the form [23]

$$\left.\begin{aligned} \underline{\underline{\varepsilon}} &= \varepsilon_1 \underline{\underline{I}} + \varepsilon_2 \underline{u}_m \underline{u}_m + i\varepsilon_3 \underline{u}_m \times \underline{\underline{I}} \\ \underline{\underline{\xi}} &= i\left(\xi_1 \underline{\underline{I}} + \xi_2 \underline{u}_m \underline{u}_m + i\xi_3 \underline{u}_m \times \underline{\underline{I}}\right) \\ \underline{\underline{\zeta}} &= -\underline{\underline{\xi}} \\ \underline{\underline{\mu}} &= \mu_1 \underline{\underline{I}} + \mu_2 \underline{u}_m \underline{u}_m + i\mu_3 \underline{u}_m \times \underline{\underline{I}} \end{aligned}\right\}, \tag{2.38}$$

$$\sigma_1 \in \mathbb{R}, \quad \sigma_2 \in \mathbb{R}, \quad \sigma_2 \in \mathbb{R}, \quad \sigma \in \{\varepsilon, \xi, \mu\}, \tag{2.39}$$

in the absence of dissipation. For dissipative Faraday chiral materials, the constitutive parameters σ_1, σ_2, and σ_3, $\sigma \in \{\varepsilon, \xi, \mu\}$, are complex valued.

Naturally occurring materials which exhibit appreciable bianisotropic effects, under normal environmental conditions, are relatively scarce. However, artificial materials with substantial bianisotropic effects may be readily realized. Such materials can be conceptualized as homogenized composite materials, arising from constituent materials which are not themselves bianisotropic (or even anisotropic in some cases) [24].

Far from being an esoteric property, bianisotropy is actually a ubiquitous one [10]. From the perspective of special relativity, isotropy is not invariant under the Lorentz transformation. For example, a material which is an isotropic dielectric material with respect to one inertial reference frame is a bianisotropic material with respect to another inertial reference frame. Furthermore, from the perspective of general relativity, free space subjected to a gravitational field is electromagnetically equivalent to a nonhomogeneous bianisotropic material [25, 26].

2.5   WAVE PROPAGATION

Let us consider wave propagation in a homogeneous bianisotropic material described by the constitutive relations (2.25), with constitutive dyadics (2.29) and (2.30). Suppose that the spatial variation of all field phasors is of the form $\exp[iq(x\cos\psi + y\sin\psi)]$ in the xy plane, with q being a complex-valued wavenumber and the angle $\psi \in [0, 2\pi)$. In particular, the field phasors are expressed as

$$\left.\begin{aligned} \underline{E}(\underline{r}) &= \underline{e}(z)\exp[iq(x\cos\psi + y\sin\psi)] \\ \underline{H}(\underline{r}) &= \underline{h}(z)\exp[iq(x\cos\psi + y\sin\psi)] \end{aligned}\right\}, \tag{2.40}$$

with auxiliary phasors

$$\left.\begin{aligned} \underline{e}(z) &= e_x(z)\underline{u}_x + e_y(z)\underline{u}_y + e_z(z)\underline{u}_z \\ \underline{h}(z) &= h_x(z)\underline{u}_x + h_y(z)\underline{u}_y + h_z(z)\underline{u}_z \end{aligned}\right\}. \tag{2.41}$$

The spatial variation along the z axis will be determined in due course. In the absence of sources, propagation is dictated by the frequency-domain Maxwell curl postulates $(2.16)_{1,2}$ with $\underline{J}_{ext}(\underline{r}) \equiv \underline{0}$.

2.5.1   MATRIX ORDINARY DIFFERENTIAL EQUATION

The combination of the constitutive relations (2.25) and the source-free counterparts of the frequency-domain Maxwell curl postulates $(2.16)_{1,2}$, together with the field phasor representation (2.40), delivers the system of four coupled ordinary differential equations

$$
\left.
\begin{aligned}
\frac{d}{dz}e_x(z) = & \; i\omega\left[\zeta_{yx}e_x(z) + \zeta_{yy}e_y(z) + \left(\zeta_{yz} + \frac{q}{\omega}\cos\psi\right)e_z(z)\right. \\
& \left. + \mu_{yx}h_x(z) + \mu_{yy}h_y(z) + \mu_{yz}h_z(z)\right] \\[4pt]
\frac{d}{dz}e_y(z) = & \; -i\omega\left[\zeta_{xx}e_x(z) + \zeta_{xy}e_y(z) + \left(\zeta_{xz} - \frac{q}{\omega}\sin\psi\right)e_z(z)\right. \\
& \left. + \mu_{xx}h_x(z) + \mu_{xy}h_y(z) + \mu_{xz}h_z(z)\right] \\[4pt]
\frac{d}{dz}h_x(z) = & \; -i\omega\left[\varepsilon_{yx}e_x(z) + \varepsilon_{yy}e_y(z) + \varepsilon_{yz}e_z(z)\right. \\
& \left. + \xi_{yx}h_x(z) + \xi_{yy}h_y(z) + \left(\xi_{yz} - \frac{q}{\omega}\cos\psi\right)h_z(z)\right] \\[4pt]
\frac{d}{dz}h_y(z) = & \; i\omega\left[\varepsilon_{xx}e_x(z) + \varepsilon_{xy}e_y(z) + \varepsilon_{xz}e_z(z)\right. \\
& \left. + \xi_{xx}h_x(z) + \xi_{xy}h_y(z) + \left(\xi_{xz} + \frac{q}{\omega}\sin\psi\right)h_z(z)\right]
\end{aligned}
\right\}, \tag{2.42}
$$

as well as the two algebraic equations

$$
\left.
\begin{aligned}
\varepsilon_{zz}e_z(z) + \xi_{zz}h_z(z) = & \; -\varepsilon_{zx}e_x(z) - \varepsilon_{zy}e_y(z) \\
& -\left(\xi_{zx} - \frac{q}{\omega}\sin\psi\right)h_x(z) - \left(\xi_{zy} + \frac{q}{\omega}\cos\psi\right)h_y(z) \\[4pt]
\zeta_{zz}e_z(z) + \mu_{zz}h_z(z) = & \; -\left(\zeta_{zx} + \frac{q}{\omega}\sin\psi\right)e_x(z) - \left(\zeta_{zy} - \frac{q}{\omega}\cos\psi\right)e_y(z) \\
& -\mu_{zx}h_x(z) - \mu_{zy}h_y(z)
\end{aligned}
\right\}. \tag{2.43}
$$

Provided that the remote possibility of

$$
\varepsilon_{zz}\mu_{zz} = \xi_{zz}\zeta_{zz} \tag{2.44}
$$

is discounted, the two algebraic equations (2.43) may be solved to obtain the following explicit expressions for the z-directed components of the auxiliary phasors:

$$
\left.
\begin{aligned}
e_z(z) &= v_{zx}^{ee}e_x(z) + v_{zy}^{ee}e_y(z) + v_{zx}^{eh}h_x(z) + v_{zy}^{eh}h_y(z) \\
h_z(z) &= v_{zx}^{he}e_x(z) + v_{zy}^{he}e_y(z) + v_{zx}^{hh}h_x(z) + v_{zy}^{hh}h_y(z)
\end{aligned}
\right\}. \tag{2.45}
$$

Herein, the coefficients

$$
\left.\begin{aligned}
v_{zx}^{ee} &= -\frac{\mu_{zz}\varepsilon_{zx} - \xi_{zz}\left[\zeta_{zx} + (q/\omega)\sin\psi\right]}{\varepsilon_{zz}\mu_{zz} - \xi_{zz}\zeta_{zz}} \\
v_{zy}^{ee} &= -\frac{\mu_{zz}\varepsilon_{zy} - \xi_{zz}\left[\zeta_{zy} - (q/\omega)\cos\psi\right]}{\varepsilon_{zz}\mu_{zz} - \xi_{zz}\zeta_{zz}} \\
v_{zx}^{eh} &= \frac{\xi_{zz}\mu_{zx} - \mu_{zz}\left[\xi_{zx} - (q/\omega)\sin\psi\right]}{\varepsilon_{zz}\mu_{zz} - \xi_{zz}\zeta_{zz}} \\
v_{zy}^{eh} &= \frac{\xi_{zz}\mu_{zy} - \mu_{zz}\left[\xi_{zy} + (q/\omega)\cos\psi\right]}{\varepsilon_{zz}\mu_{zz} - \xi_{zz}\zeta_{zz}} \\
v_{zx}^{he} &= \frac{\zeta_{zz}\varepsilon_{zx} - \varepsilon_{zz}\left[\zeta_{zx} + (q/\omega)\sin\psi\right]}{\varepsilon_{zz}\mu_{zz} - \xi_{zz}\zeta_{zz}} \\
v_{zy}^{he} &= \frac{\zeta_{zz}\varepsilon_{zy} - \varepsilon_{zz}\left[\zeta_{zy} - (q/\omega)\cos\psi\right]}{\varepsilon_{zz}\mu_{zz} - \xi_{zz}\zeta_{zz}} \\
v_{zx}^{hh} &= -\frac{\varepsilon_{zz}\mu_{zx} - \zeta_{zz}\left[\xi_{zx} - (q/\omega)\sin\psi\right]}{\varepsilon_{zz}\mu_{zz} - \xi_{zz}\zeta_{zz}} \\
v_{zy}^{hh} &= -\frac{\varepsilon_{zz}\mu_{zy} - \zeta_{zz}\left[\xi_{zy} + (q/\omega)\cos\psi\right]}{\varepsilon_{zz}\mu_{zz} - \xi_{zz}\zeta_{zz}}
\end{aligned}\right\} . \tag{2.46}
$$

The explicit expressions for $e_z(z)$ and $h_z(z)$ given in (2.45) may be substituted into the system of ordinary differential equations (2.42) to obtain the 4×4 matrix ordinary differential equation

$$
\frac{d}{dz}\left[\underline{f}(z)\right] = i\left[\underline{\underline{P}}\right] \cdot \left[\underline{f}(z)\right] . \tag{2.47}
$$

Herein, the 4-column vector

$$
\left[\underline{f}(z)\right] = \begin{bmatrix} e_x(z) \\ e_y(z) \\ h_x(z) \\ h_y(z) \end{bmatrix} , \tag{2.48}
$$

while the 4×4 matrix

$$
\left[\underline{\underline{P}}\right] = \omega \left(\left[\begin{array}{cccc} \zeta_{yx} & \zeta_{yy} & \mu_{yx} & \mu_{yy} \\ -\zeta_{xx} & -\zeta_{xy} & -\mu_{xx} & -\mu_{xy} \\ -\varepsilon_{yx} & -\varepsilon_{yy} & -\xi_{yx} & -\xi_{yy} \\ \varepsilon_{xx} & \varepsilon_{xy} & \xi_{xx} & \xi_{xy} \end{array} \right] + \right.
$$

$$
\left[\begin{array}{cccc} \zeta_{yz} + \dfrac{q}{\omega}\cos\psi & 0 & 0 & 0 \\ 0 & -\zeta_{xz} + \dfrac{q}{\omega}\sin\psi & 0 & 0 \\ 0 & 0 & -\varepsilon_{yz} & 0 \\ 0 & 0 & 0 & \varepsilon_{xz} \end{array} \right] \cdot \left[\underline{\underline{J}}\right] \cdot \left[\begin{array}{cccc} v_{zx}^{ee} & 0 & 0 & 0 \\ 0 & v_{zy}^{ee} & 0 & 0 \\ 0 & 0 & v_{zx}^{eh} & 0 \\ 0 & 0 & 0 & v_{zy}^{eh} \end{array} \right] +
$$

$$
\left. \left[\begin{array}{cccc} \mu_{yz} & 0 & 0 & 0 \\ 0 & -\mu_{xz} & 0 & 0 \\ 0 & 0 & -\xi_{yz} + \dfrac{q}{\omega}\cos\psi & 0 \\ 0 & 0 & 0 & \xi_{xz} + \dfrac{q}{\omega}\sin\psi \end{array} \right] \cdot \left[\underline{\underline{J}}\right] \cdot \left[\begin{array}{cccc} v_{zx}^{he} & 0 & 0 & 0 \\ 0 & v_{zy}^{he} & 0 & 0 \\ 0 & 0 & v_{zx}^{hh} & 0 \\ 0 & 0 & 0 & v_{zy}^{hh} \end{array} \right] \right)
$$

$$(2.49)$$

is expressed using the 4×4 matrix

$$
\left[\underline{\underline{J}}\right] = \left[\begin{array}{cccc} 1 & 1 & 1 & 1 \\ 1 & 1 & 1 & 1 \\ 1 & 1 & 1 & 1 \\ 1 & 1 & 1 & 1 \end{array} \right]. \tag{2.50}
$$

2.5.2 TRANSFER MATRIX

Since the bianisotropic material under consideration is homogeneous, the matrix $\left[\underline{\underline{P}}\right]$ on the right side of (2.47) does not depend upon z (or indeed x and y). Accordingly, the solution of (2.47) is given as [27]

$$
\left[\underline{f}(z)\right] = \exp\left\{i\left[\underline{\underline{P}}\right]z\right\} \cdot \left[\underline{f}(0)\right] \tag{2.51}
$$

when the boundary value of $\left[\underline{f}(z)\right]$ is specified at $z = 0$. The matrizant

$$
\left[\underline{\underline{M}}(z)\right] = \exp\left\{i\left[\underline{\underline{P}}\right]z\right\} \tag{2.52}
$$

satisfies the matrix ordinary differential equation

$$
\frac{d}{dz}\left[\underline{\underline{M}}(z)\right] = i\left[\underline{\underline{P}}\right] \cdot \left[\underline{\underline{M}}(z)\right] \tag{2.53}
$$

subject to the boundary condition

$$
\left[\underline{\underline{M}}(0)\right] = \left[\underline{\underline{I}}\right], \tag{2.54}
$$

where $\left[\underline{\underline{I}}\right]$ is the 4×4 identity matrix.

More generally, if the boundary value of $\left[\underline{f}(z)\right]$ is specified at $z = z_a$ then the solution of (2.47) may be expressed as

$$\left[\underline{f}(z)\right] = \exp\left\{i\left[\underline{\underline{P}}\right](z - z_a)\right\} \cdot \left[\underline{f}(z_a)\right], \qquad (2.55)$$

i.e.,

$$\left[\underline{f}(z)\right] = \left[\underline{\underline{M}}(z - z_a)\right] \cdot \left[\underline{f}(z_a)\right]. \qquad (2.56)$$

If $z = z_a$ and $z = z_b$ are the two faces of a layer, then the matrix

$$\left[\underline{\underline{M}}\right] = \left[\underline{\underline{M}}(z_b - z_a)\right] \qquad (2.57)$$

is called the *transfer matrix* because it relates the value of $\left[\underline{f}(z)\right]$ at the face $z = z_b$ to the value of $\left[\underline{f}(z)\right]$ at the face $z = z_a$ as follows:

$$\left[\underline{f}(z_b)\right] = \left[\underline{\underline{M}}\right] \cdot \left[\underline{f}(z_a)\right]. \qquad (2.58)$$

Significantly, the transfer matrix $\left[\underline{\underline{M}}\right]$ emerges directly from the frequency-domain Maxwell curl postulates without taking recourse to differentiation with respect to z.

The matrix exponential term on the right sides of (2.51) and (2.55) is defined as the sum

$$\left[\underline{\underline{M}}(z)\right] = \sum_{q=0}^{\infty} \frac{i^q}{q!} \left[\underline{\underline{P}}\right]^q z^q, \qquad (2.59)$$

which may be computed using standard software packages; e.g., Mathematica™ uses the command MatrixExp, and Matlab™ the command expm.

2.5.3   DIAGONALIZATION

An alternative, often neater, way of dealing with the matrix exponential term on the right sides of (2.51) and (2.55) involves diagonalizing the 4×4 matrix $\left[\underline{\underline{P}}\right]$ as follows. This option is available only when $\left[\underline{\underline{P}}\right]$ has four linearly independent eigenvectors [28]. Although this case is most likely true in practice, exceptional cases, wherein $\left[\underline{\underline{P}}\right]$ is not diagonalizable, have been found in instances of surface waves with unusual localization characteristics [29–31].

Let the m-th eigenvalue of $\left[\underline{\underline{P}}\right]$ be represented by the scalar g_m, $m \in [1, 4]$, and the 4-column vector

$$[\underline{v}]^{(m)} = \begin{bmatrix} v_{1m} \\ v_{2m} \\ v_{3m} \\ v_{4m} \end{bmatrix} \tag{2.60}$$

be the corresponding eigenvector. Then $\left[\underline{\underline{P}}\right]$ can be factored as

$$\left[\underline{\underline{P}}\right] = \left[\underline{\underline{V}}\right] \cdot \left[\underline{\underline{G}}\right] \cdot \left[\underline{\underline{V}}\right]^{-1} , \tag{2.61}$$

with the matrix

$$\left[\underline{\underline{V}}\right] = \begin{bmatrix} v_{11} & v_{12} & v_{13} & v_{14} \\ v_{21} & v_{22} & v_{23} & v_{24} \\ v_{31} & v_{32} & v_{33} & v_{34} \\ v_{41} & v_{42} & v_{43} & v_{44} \end{bmatrix} \tag{2.62}$$

comprising the eigenvectors and the diagonal matrix

$$\left[\underline{\underline{G}}\right] = \begin{bmatrix} g_1 & 0 & 0 & 0 \\ 0 & g_2 & 0 & 0 \\ 0 & 0 & g_3 & 0 \\ 0 & 0 & 0 & g_4 \end{bmatrix} \tag{2.63}$$

comprising the eigenvalues of $\left[\underline{\underline{P}}\right]$. It is often convenient to implement the normalization

$$|v_{1m}|^2 + |v_{2m}|^2 = 1, \quad m \in [1, 4] , \tag{2.64}$$

which makes the magnitude of the electric field phasor associated with each eigenvector close to unity.

The diagonalization of $\left[\underline{\underline{P}}\right]$ through (2.61) into $\left[\underline{\underline{G}}\right]$ allows the matrix exponential term on the right sides of (2.51) and (2.55) to be expressed as

$$\left[\underline{\underline{M}}(z)\right] = \left[\underline{\underline{V}}\right] \cdot \exp\left\{i \left[\underline{\underline{G}}\right] z\right\} \cdot \left[\underline{\underline{V}}\right]^{-1} , \tag{2.65}$$

wherein

$$\exp\left\{i \left[\underline{\underline{G}}\right] z\right\} = \begin{bmatrix} \exp(i g_1 z) & 0 & 0 & 0 \\ 0 & \exp(i g_2 z) & 0 & 0 \\ 0 & 0 & \exp(i g_3 z) & 0 \\ 0 & 0 & 0 & \exp(i g_4 z) \end{bmatrix} . \tag{2.66}$$

And thus the solution (2.55) may be reformulated as

$$\left[\underline{f}(z)\right] = \left[\underline{\underline{V}}\right] \bullet \exp\left\{i\left[\underline{\underline{G}}\right](z - z_a)\right\} \bullet \left[\underline{\underline{V}}\right]^{-1} \bullet \left[\underline{f}(z_a)\right], \tag{2.67}$$

which is tractable for both analysis and computation.

Notice that (2.51) and (2.55) are solutions of the matrix ordinary differential equation (2.47) whether or not $\left[\underline{\underline{P}}\right]$ is diagonalizable, but (2.67) is only a solution when $\left[\underline{\underline{P}}\right]$ is diagonalizable.

2.5.4   EIGENMODES

The eigenanalysis of the 4×4 matrix $\left[\underline{\underline{P}}\right]$ facilitates an eigenmodal decomposition of the field phasors, because the phasors may be represented as

$$\left.\begin{aligned} \underline{E}(\underline{r}) &= \sum_{m=1}^{4} C_m \, \underline{E}^{(m)}(\underline{r}) \\ \underline{H}(\underline{r}) &= \sum_{m=1}^{4} C_m \, \underline{H}^{(m)}(\underline{r}) \end{aligned}\right\}, \tag{2.68}$$

wherein the m-th eigenmode is identified by

$$\left.\begin{aligned} \underline{E}^{(m)}(\underline{r}) &= \left[v_{1m}\,\underline{u}_x + v_{2m}\,\underline{u}_y + \left(v_{zx}^{ee}\,v_{1m} + v_{zy}^{ee}\,v_{2m} + v_{zx}^{eh}\,v_{3m} + v_{zy}^{eh}\,v_{4m}\right)\underline{u}_z\right] \\ &\quad \times \exp\left[iq(x\cos\psi + y\sin\psi)\right]\exp(ig_m z) \\ \underline{H}^{(m)}(\underline{r}) &= \left[v_{3m}\,\underline{u}_x + v_{4m}\,\underline{u}_y + \left(v_{zx}^{he}\,v_{1m} + v_{zy}^{he}\,v_{2m} + v_{zx}^{hh}\,v_{3m} + v_{zy}^{hh}\,v_{4m}\right)\underline{u}_z\right] \\ &\quad \times \exp\left[iq(x\cos\psi + y\sin\psi)\right]\exp(ig_m z) \end{aligned}\right\} \tag{2.69}$$

and the coefficients C_m, $m \in [1, 4]$, can be obtained from

$$\begin{bmatrix} C_1 \\ C_2 \\ C_3 \\ C_4 \end{bmatrix} = \left[\underline{\underline{V}}\right]^{-1} \bullet \left[\underline{f}(0)\right]. \tag{2.70}$$

The rate at which energy is transported by the m-th eigenmode is delivered by the corresponding time-averaged Poynting vector [8]

$$\underline{P}^{(m)}(\underline{r}) = (1/2)\,\mathrm{Re}\left\{\underline{E}^{(m)}(\underline{r}) \times \left[\underline{H}^{(m)}(\underline{r})\right]^*\right\}, \quad m \in [1, 4]. \tag{2.71}$$

Two eigenmodes transport energy along the $+z$ axis and two along the $-z$ axis. The former are labeled $m = 1$ and $m = 2$, the latter $m = 3$ and $m = 4$. In order to implement this eigenmodal classification, the z-directed component of the time-averaged Poynting vector

$$P_z^{(m)}(\underline{r}) = (1/2)\,\mathrm{Re}\,\{v_{1m}v_{4m}^* - v_{2m}v_{3m}^*\}$$
$$\times \exp\{-2\,[\mathrm{Im}\,\{q\}\,(x\cos\psi + y\sin\psi) + \mathrm{Im}\,\{g_m\}\,z]\}\,,\quad m \in [1,4]\,,\quad (2.72)$$

is utilized, with

$$\left.\begin{array}{l} P_z^{(1)}(\underline{r}) > 0 \\[4pt] P_z^{(2)}(\underline{r}) > 0 \end{array}\right\} \qquad (2.73)$$

and

$$\left.\begin{array}{l} P_z^{(3)}(\underline{r}) < 0 \\[4pt] P_z^{(4)}(\underline{r}) < 0 \end{array}\right\}. \qquad (2.74)$$

The 4-column vector $\left[\underline{f}(z)\right]$ may be partitioned accordingly as

$$\left[\underline{f}(z)\right] = \left[\underline{f}^+(z)\right] + \left[\underline{f}^-(z)\right], \qquad (2.75)$$

where

$$\left[\underline{f}^+(z)\right] = \begin{bmatrix} v_{11} & v_{12} \\ v_{21} & v_{22} \\ v_{31} & v_{32} \\ v_{41} & v_{42} \end{bmatrix} \cdot \begin{bmatrix} \exp(ig_1z) & 0 \\ 0 & \exp(ig_2z) \end{bmatrix} \cdot \begin{bmatrix} C_1 \\ C_2 \end{bmatrix} \qquad (2.76)$$

and

$$\left[\underline{f}^-(z)\right] = \begin{bmatrix} v_{13} & v_{14} \\ v_{23} & v_{24} \\ v_{33} & v_{34} \\ v_{43} & v_{44} \end{bmatrix} \cdot \begin{bmatrix} \exp(ig_3z) & 0 \\ 0 & \exp(ig_4z) \end{bmatrix} \cdot \begin{bmatrix} C_3 \\ C_4 \end{bmatrix}. \qquad (2.77)$$

A contradiction to this arrangement of two eigenmodes of each type has not been reported as yet.

The electromagnetic fields cannot grow in magnitude along the direction of energy transportation in a passive material. Accordingly, the eigenmodes labeled $m \in [1,2]$ are required to satisfy the constraint

$$\left.\begin{array}{l} \mathrm{Im}\,\{g_1\} \geq 0 \\[4pt] \mathrm{Im}\,\{g_2\} \geq 0 \end{array}\right\}, \qquad (2.78)$$

whereas the eigenmodes labeled $m \in [3,4]$ are required to satisfy the constraint

$$\left.\begin{array}{l} \mathrm{Im}\,\{g_3\} \leq 0 \\[4pt] \mathrm{Im}\,\{g_4\} \leq 0 \end{array}\right\}. \qquad (2.79)$$

The inequalities (2.78) and (2.79) are consistent with the inequalities (2.73) and (2.74).

2.6   REFERENCES

[1] Jackson, J. D. 1999. *Classical Electrodynamics*, 3rd ed. (New York, NY, Wiley). 12, 13

[2] Buchwald, J. Z. 1985. *From Maxwell to Microphysics: Aspects of Electromagnetic Theory in the Last Quarter of the Nineteenth Century* (Chicago, IL, University of Chicago Press). 12

[3] Weiglhofer, W. S. 2003. Constitutive characterization of simple and complex mediums, in *Introduction to Complex Mediums for Optics and Electromagnetics*, Eds., W. S. Weiglhofer and A. Lakhtakia, pages 27–61 (Bellingham, WA, SPIE Press). 15, 17

[4] Chern, R.-L. 2013. Spatial dispersion and nonlocal effective permittivity for periodic layered metamaterials, *Opt. Express*, 21:16514–16527. DOI: 10.1364/oe.21.016514. 15

[5] Arfken, G. B. and Weber, H. J. 1995. *Mathematical Methods for Physicists*, 4th ed. (London, Academic Press). 16

[6] Ashcroft, N. W. and Mermin, N. D. 1976. *Solid State Physics* (Phildelphia, PA, Saunders). 17

[7] Hilgevoord, J. 1962. *Dispersion Relations and Causal Descriptions* (Amsterdam, The Netherlands, North-Holland). 17

[8] Chen, H. C. 1983. *Theory of Electromagnetic Waves* (New York, McGraw-Hill). 17, 21, 28

[9] Lakhtakia, A. and Weiglhofer, W. S. 2000. Electromagnetic waves in a material with simultaneous mirror-conjugated and racemic chirality characteristics, *Electromagnetics*, 20:481–488 (the first negative sign on the right side of Eq. (26) of this paper should be replaced by a positive sign). DOI: 10.1080/027263400750068523. 17

[10] Mackay, T. G. and Lakhtakia, A. 2019. *Electromagnetic Anisotropy and Bianisotropy: A Field Guide*, 2nd ed. (Singapore, World Scientific). DOI: 10.1142/11351. 18, 20, 21, 22

[11] Lakhtakia, A. 1994. *Beltrami Fields in Chiral Media* (Singapore, World Scientific). DOI: 10.1142/2031. 18

[12] Lakhtakia, A., Ed., 1990. *Selected Papers on Natural Optical Activity* (Bellingham, WA, SPIE Optical Engineering Press). 18

[13] Hasan, M. Z. and Kane, C. L. 2010. Topological insulators, *Rev. Mod. Phys.*, 82:3045–3067. DOI: 10.1103/RevModPhys.82.3045. 19

[14] Lakhtakia, A. and Mackay, T. G. 2016. Classical electromagnetic model of surface states in topological insulators, *J. Nanophoton.*, 10:033004. DOI: 10.1117/1.jnp.10.033004. 19

[15] Wakaki, M. 2012. *Optical Materials and Applications* (London, CRC Press). DOI: 10.1201/b12496. 19

[16] Dmitriev, V. 1999. Group theoretical approach to complex and bianisotropic media description, *Eur. Phys. J. Appl. Phys.*, 6:49–55. DOI: 10.1051/epjap:1999151. 20

[17] Litvin, D. B. 2003. Point group symmetries, in *Introduction to Complex Mediums for Optics and Electromagnetics*, Eds., W. S. Weiglhofer and A. Lakhtakia, pages 79–102 (Bellingham, WA, SPIE Press). DOI: 10.1117/3.504610. 20

[18] Lifshitz, R. 2005. Magnetic point groups and space groups, *Encyclopedia of Condensed Matter Physics*, 3rd ed., Eds., F. Bassani, G. L. Liedl, and P. Wyder, pages 219–226 (Oxford, UK, Elsevier). 20

[19] Born, M. and Wolf, E. 1999. *Principles of Optics*, 7th ed. (Cambridge, UK, Cambridge University Press). 21

[20] Weiglhofer, W. S. and Lakhtakia, A. 1999. On electromagnetic waves in biaxial bianisotropic media, *Electromagnetics*, 19:351–362. DOI: 10.1080/02726349908908652. 21

[21] Mackay, T. G. and Weiglhofer, W. S. 2000. Homogenization of biaxial composite materials: Dissipative anisotropic properties, *J. Opt. A: Pure Appl. Opt.*, 2:426–432. DOI: 10.1088/1464-4258/2/5/313. 21

[22] Lax, B. and Button, K. J. 1962. *Microwave Ferrites and Ferrimagnetics* (New York, McGraw–Hill). 21

[23] Weiglhofer, W. S. and Lakhtakia, A. 1998. The correct constitutive relations of chiroplasmas and chiroferrites, *Microw. Opt. Technol. Lett.*, 17:405–408. DOI: 10.1002/(sici)1098-2760(19980420)17:6%3C405::aid-mop17%3E3.0.co;2-z. 22

[24] Mackay, T. G. and Lakhtakia, A. 2015. *Modern Analytical Electromagnetic Homogenization* (San Rafael, CA, Morgan & Claypool), IOP Concise Physics. DOI: 10.1088/978-1-6270-5427-0. 22

[25] Plébanski, J. 1960. Electromagnetic waves in gravitational fields, *Phys. Rev.*, 118:1396–1408. DOI: 10.1103/physrev.118.1396. 22

[26] Schleich, W. and Scully, M. O. 1984. General relativity and modern optics, *New Trends in Atomic Physics*, Eds., G. Grynberg and R. Stora, pages 995–1124 (Amsterdam, The Netherlands: Elsevier). 22

[27] Hochstadt, H. 1975. *Differential Equations: A Modern Approach* (New York, Dover Press). 25

[28] Kreyszig, E. 1988. *Advanced Engineering Mathematics*, 6th ed. (New York, Wiley). 26

[29] Mackay, T. G., Zhou, C., and Lakhtakia, A. 2019. Dyakonov–Voigt surface waves, *Proc. R. Soc. A*, 475:20190317. DOI: 10.1098/rspa.2019.0317. 26

[30] Zhou, C., Mackay, T. G., and Lakhtakia, A. 2019. On Dyakonov–Voigt surface waves guided by the planar interface of dissipative materials, *J. Opt. Soc. Am. B*, 36:3218–3225. DOI: 10.1364/JOSAB.36.003218. 26

[31] Zhou, C., Mackay, T. G., and Lakhtakia, A. 2019. Surface-plasmon-polariton wave propagation supported by anisotropic materials: Multiple modes and mixed exponential and linear localization characteristics, *Phys. Rev. A*, 100:033809. DOI: 10.1103/physreva.100.033809. 26

CHAPTER 3

Bianisotropic Slab with Planar Interfaces

The TMM is developed in this chapter for the planewave-response characteristics of a multi-layered slab comprising three layers of different materials. Each constituent layer is made of a different homogeneous bianisotropic material [1] and has planar surfaces. This arrangement is appropriate for the analysis of surface waves excited in prism-coupled configurations [2, 3], such as the Turbadar–Kretschmann–Raether configuration [4–6] and the Turbadar–Otto configuration [7]. It is also applicable to reflection-transmission studies involving a multilayered slab with one or more constituent layers which are made of bianisotropic (or anisotropic) materials that are nonhomogeneous in the thickness direction.

Nonhomogeneity in the thickness direction may be accommodated by a piecewise-uniform approximation [3, 8]. Such applications arise when certain sculptured thin films [8] are incorporated in the multilayered slab as well as in the analysis of Tamm surface waves [9, 10] and Dyakonov–Tamm surface waves [11, 12].

3.1 PRELIMINARIES

Consider a multilayered slab comprising three constituent layers, labeled $\mathcal{A}$, $\mathcal{B}$, and $\mathcal{C}$, each of infinite extent in the xy plane. All surfaces and interfaces are all planar. Layer $\mathcal{A}$ has thickness $d_{\mathcal{A}}$, layer $\mathcal{B}$ has thickness $d_{\mathcal{B}}$, and layer $\mathcal{C}$ has thickness $d_{\mathcal{C}}$. Each layer is made of a different homogeneous bianisotropic material. In light of (2.25), the constitutive relations of the material of each layer are specified as

$$\left. \begin{aligned} \underline{D}(\underline{r}) &= \underline{\underline{\varepsilon}}^{\mathcal{A}} \cdot \underline{E}(\underline{r}) + \underline{\underline{\xi}}^{\mathcal{A}} \cdot \underline{H}(\underline{r}) \\ \underline{B}(\underline{r}) &= \underline{\underline{\zeta}}^{\mathcal{A}} \cdot \underline{E}(\underline{r}) + \underline{\underline{\mu}}^{\mathcal{A}} \cdot \underline{H}(\underline{r}) \end{aligned} \right\}, \quad 0 < z < d_{\mathcal{A}}, \tag{3.1}$$

$$\left. \begin{aligned} \underline{D}(\underline{r}) &= \underline{\underline{\varepsilon}}^{\mathcal{B}} \cdot \underline{E}(\underline{r}) + \underline{\underline{\xi}}^{\mathcal{B}} \cdot \underline{H}(\underline{r}) \\ \underline{B}(\underline{r}) &= \underline{\underline{\zeta}}^{\mathcal{B}} \cdot \underline{E}(\underline{r}) + \underline{\underline{\mu}}^{\mathcal{B}} \cdot \underline{H}(\underline{r}) \end{aligned} \right\}, \quad d_{\mathcal{A}} < z < d_{\mathcal{A}} + d_{\mathcal{B}}, \tag{3.2}$$

and

$$\left. \begin{aligned} \underline{D}(\underline{r}) &= \underline{\underline{\varepsilon}}^{\mathcal{C}} \cdot \underline{E}(\underline{r}) + \underline{\underline{\xi}}^{\mathcal{C}} \cdot \underline{H}(\underline{r}) \\ \underline{B}(\underline{r}) &= \underline{\underline{\zeta}}^{\mathcal{C}} \cdot \underline{E}(\underline{r}) + \underline{\underline{\mu}}^{\mathcal{C}} \cdot \underline{H}(\underline{r}) \end{aligned} \right\}, \quad d_{\mathcal{A}} + d_{\mathcal{B}} < z < d, \tag{3.3}$$

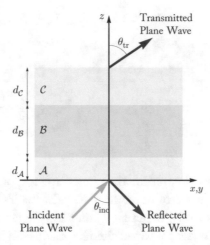

Figure 3.1: Schematic representation of the reflection and transmission of a plane wave by a multilayered slab comprising layers $\mathcal{A}$, $\mathcal{B}$, and $\mathcal{C}$. The layers $\mathcal{A}$ and $\mathcal{B}$ have a planar interface $z = d_{\mathcal{A}}$, and the layers $\mathcal{B}$ and $\mathcal{C}$ have a planar interface $z = d_{\mathcal{A}} + d_{\mathcal{B}}$. The angle of incidence $\theta_{\text{inc}} \in [0, \pi/2)$, but the angle θ_{tr} can be either real valued or complex valued.

with $d = d_{\mathcal{A}} + d_{\mathcal{B}} + d_{\mathcal{C}}$. The constitutive dyadics in (3.1)–(3.3) have the general form given in (2.29) and (2.30), with the symbols ε, ξ, ζ, and μ therein replaced by, respectively, $\varepsilon^{\gamma}, \xi^{\gamma}, \zeta^{\gamma}$, and μ^{γ}, $\gamma \in \{\mathcal{A}, \mathcal{B}, \mathcal{C}\}$.

A homogeneous isotropic dielectric material fills the half-space $z < 0$. Its refractive index n_1 is taken to be real valued and positive. Likewise, the half-space $z > d$ is filled with a homogeneous isotropic dielectric material characterized by the refractive index n_2 which is also real valued and positive. The dependences of the constitutive parameters of all five materials on the angular frequency ω are implicit in the remainder of this chapter.

3.2   INCIDENT, REFLECTED, AND TRANSMITTED PLANE WAVES

The reflection-transmission problem is illustrated schematically in Fig. 3.1. A plane wave propagating in the half-space $z < 0$ is incident on the plane $z = 0$. The direction of propagation of the incident plane is inclined at an angle $\theta_{\text{inc}} \in [0, \pi/2)$ with respect to the z axis and at an angle $\psi \in [0, 2\pi)$ with respect to the x axis in the xy plane. The polarization state of this incident plane wave is arbitrary, but it is convenient to express the field phasors associated with the incident plane wave in terms of s- and p-polarized components per

$$\left.\begin{aligned}
\underline{E}_{\text{inc}}(\underline{r}) &= \left(a_s\,\underline{s} + a_p\,\underline{p}_{\text{inc}}\right) \exp\left\{i\left[q\left(x\cos\psi + y\sin\psi\right) + k_0 n_1 z\cos\theta_{\text{inc}}\right]\right\} \\
\underline{H}_{\text{inc}}(\underline{r}) &= n_1\eta_0^{-1}\left(a_s\,\underline{p}_{\text{inc}} - a_p\,\underline{s}\right) \exp\left\{i\left[q\left(x\cos\psi + y\sin\psi\right) + k_0 n_1 z\cos\theta_{\text{inc}}\right]\right\}
\end{aligned}\right\},$$

$$z < 0. \tag{3.4}$$

In these equations and hereafter, the free-space wavenumber is denoted by $k_0 = \omega\sqrt{\varepsilon_0\mu_0}$, the intrinsic impedance of free space is denoted by $\eta_0 = \sqrt{\mu_0/\varepsilon_0}$, and the real-valued wavenumber

$$q = k_0 n_1 \sin\theta_{\text{inc}}. \tag{3.5}$$

Denoting, respectively, the amplitudes of the s- and the p-polarized components of the incident plane wave in (3.4), the parameters a_s and a_p are presumed to be known, whereas the unit vectors

$$\left.\begin{aligned}
\underline{s} &= -\underline{u}_x \sin\psi + \underline{u}_y \cos\psi \\
\underline{p}_{\text{inc}} &= -\left(\underline{u}_x \cos\psi + \underline{u}_y \sin\psi\right)\cos\theta_{\text{inc}} + \underline{u}_z \sin\theta_{\text{inc}}
\end{aligned}\right\}. \tag{3.6}$$

In view of (3.4), the corresponding reflected field phasors are given as

$$\left.\begin{aligned}
\underline{E}_{\text{ref}}(\underline{r}) &= \left(r_s\,\underline{s} + r_p\,\underline{p}_{\text{ref}}\right) \exp\left\{i\left[q\left(x\cos\psi + y\sin\psi\right) - k_0 n_1 z\cos\theta_{\text{inc}}\right]\right\} \\
\underline{H}_{\text{ref}}(\underline{r}) &= n_1\eta_0^{-1}\left(r_s\,\underline{p}_{\text{ref}} - r_p\,\underline{s}\right) \exp\left\{i\left[q\left(x\cos\psi + y\sin\psi\right) - k_0 n_1 z\cos\theta_{\text{inc}}\right]\right\}
\end{aligned}\right\},$$

$$z < 0, \tag{3.7}$$

and the corresponding transmitted field phasors given as

$$\left.\begin{aligned}
\underline{E}_{\text{tr}}(\underline{r}) &= \left(t_s\,\underline{s} + t_p\,\underline{p}_{\text{tr}}\right) \exp\left\{i\left[q\left(x\cos\psi + y\sin\psi\right) + k_0 n_2(z - d)\cos\theta_{\text{tr}}\right]\right\} \\
\underline{H}_{\text{tr}}(\underline{r}) &= n_2\eta_0^{-1}\left(t_s\,\underline{p}_{\text{tr}} - t_p\,\underline{s}\right) \exp\left\{i\left[q\left(x\cos\psi + y\sin\psi\right) + k_0 n_2(z - d)\cos\theta_{\text{tr}}\right]\right\}
\end{aligned}\right\},$$

$$z > d, \tag{3.8}$$

with the unit vectors

$$\left.\begin{aligned}
\underline{p}_{\text{ref}} &= \left(\underline{u}_x \cos\psi + \underline{u}_y \sin\psi\right)\cos\theta_{\text{inc}} + \underline{u}_z \sin\theta_{\text{inc}} \\
\underline{p}_{\text{tr}} &= -\left(\underline{u}_x \cos\psi + \underline{u}_y \sin\psi\right)\cos\theta_{\text{tr}} + \underline{u}_z \sin\theta_{\text{tr}}
\end{aligned}\right\}, \tag{3.9}$$

The angles θ_{inc} and θ_{tr} are related via the formulas

$$\left.\begin{aligned}
\sin\theta_{\text{tr}} &= \frac{n_1}{n_2}\sin\theta_{\text{inc}} \\
\cos\theta_{\text{tr}} &= +\sqrt{1 - \sin^2\theta_{\text{tr}}}
\end{aligned}\right\}. \tag{3.10}$$

When $n_1 > n_2$, it is possible for $\sin\theta_{\mathrm{tr}}$ to exceed unity, in which event $\cos\theta_{\mathrm{tr}}$ is complex valued and the transmitted plane wave is evanescent.

The linear reflection amplitudes r_{s} and r_{p} in (3.7), and the linear transmission amplitudes t_{s} and t_{p} in (3.8), are presumed to be unknown, but these can determined by solving the corresponding boundary-value problem, as described next.

3.3   SOLUTION OF BOUNDARY-VALUE PROBLEM

Since the field phasors have the form given in (2.40) for all $z \in (-\infty, \infty)$, the column vector $\left[\underline{f}(z)\right]$ defined in (2.48) satisfies the matrix ordinary differential equations

$$
\frac{d}{dz}\left[\underline{f}(z)\right] = \begin{cases} i\left[\underline{\underline{P}}\right]^{\mathcal{A}} \cdot \left[\underline{f}(z)\right], & 0 < z < d_{\mathcal{A}}, \\ i\left[\underline{\underline{P}}\right]^{\mathcal{B}} \cdot \left[\underline{f}(z)\right], & d_{\mathcal{A}} < z < d_{\mathcal{A}} + d_{\mathcal{B}}, \\ i\left[\underline{\underline{P}}\right]^{\mathcal{C}} \cdot \left[\underline{f}(z)\right], & d_{\mathcal{A}} + d_{\mathcal{B}} < z < d, \end{cases} \tag{3.11}
$$

in the multilayered slab. The matrixes $\left[\underline{\underline{P}}\right]^{\gamma}$, $\gamma \in \{\mathcal{A}, \mathcal{B}, \mathcal{C}\}$, in (3.11) are defined as in (2.49), but with the symbols $\varepsilon, \xi, \zeta,$ and μ therein replaced, respectively, by $\varepsilon^{\gamma}, \xi^{\gamma}, \zeta^{\gamma},$ and μ^{γ}.

In view of the solutions (2.51) and (2.55), the relations

$$
\left.\begin{aligned} \left[\underline{f}(d_{\mathcal{A}}-)\right] &= \left[\underline{\underline{M}}\right]^{\mathcal{A}} \cdot \left[\underline{f}(0+)\right] \\ \left[\underline{f}(\{d_{\mathcal{A}} + d_{\mathcal{B}}\}-)\right] &= \left[\underline{\underline{M}}\right]^{\mathcal{B}} \cdot \left[\underline{f}(d_{\mathcal{A}}+)\right] \\ \left[\underline{f}(d-)\right] &= \left[\underline{\underline{M}}\right]^{\mathcal{C}} \cdot \left[\underline{f}(\{d_{\mathcal{A}} + d_{\mathcal{B}}\}+)\right] \end{aligned}\right\} \tag{3.12}
$$

follow from (3.11). Herein, $\left[\underline{\underline{M}}\right]^{\mathcal{A}}$, $\left[\underline{\underline{M}}\right]^{\mathcal{B}}$, and $\left[\underline{\underline{M}}\right]^{\mathcal{C}}$ are 4×4 *transfer matrixes* which are expressed in terms of matrix exponentials as

$$
\left[\underline{\underline{M}}\right]^{\gamma} = \exp\left\{i\left[\underline{\underline{P}}\right]^{\gamma} d_{\gamma}\right\}, \qquad \gamma \in \{\mathcal{A}, \mathcal{B}, \mathcal{C}\}. \tag{3.13}
$$

Now we turn to the boundary values of the tangential components of the auxiliary phasors. Whereas

$$
\left[\underline{f}(0-)\right] = \left[\underline{\underline{K}}\right]_{\mathrm{inc}} \cdot \begin{bmatrix} a_{\mathrm{s}} \\ a_{\mathrm{p}} \\ r_{\mathrm{s}} \\ r_{\mathrm{p}} \end{bmatrix} \tag{3.14}
$$

emerges immediately below the layer $\mathcal{A}$ from (3.4) and (3.7), immediately above the layer $\mathcal{C}$ we have

$$\left[\underline{f}(d+)\right] = \left[\underline{\underline{K}}\right]_{\mathrm{tr}} \cdot \begin{bmatrix} t_{\mathrm{s}} \\ t_{\mathrm{p}} \\ 0 \\ 0 \end{bmatrix} \tag{3.15}$$

by virtue of (3.8). In these equations, the 4×4 matrixes

$$\left[\underline{\underline{K}}\right]_{\mathrm{inc}} = \begin{bmatrix} -\sin\psi & -\cos\psi\cos\theta_{\mathrm{inc}} & -\sin\psi & \cos\psi\cos\theta_{\mathrm{inc}} \\ \cos\psi & -\sin\psi\cos\theta_{\mathrm{inc}} & \cos\psi & \sin\psi\cos\theta_{\mathrm{inc}} \\ -\left(\dfrac{n_1}{\eta_0}\right)\cos\psi\cos\theta_{\mathrm{inc}} & \left(\dfrac{n_1}{\eta_0}\right)\sin\psi & \left(\dfrac{n_1}{\eta_0}\right)\cos\psi\cos\theta_{\mathrm{inc}} & \left(\dfrac{n_1}{\eta_0}\right)\sin\psi \\ -\left(\dfrac{n_1}{\eta_0}\right)\sin\psi\cos\theta_{\mathrm{inc}} & -\left(\dfrac{n_1}{\eta_0}\right)\cos\psi & \left(\dfrac{n_1}{\eta_0}\right)\sin\psi\cos\theta_{\mathrm{inc}} & -\left(\dfrac{n_1}{\eta_0}\right)\cos\psi \end{bmatrix} \tag{3.16}$$

and

$$\left[\underline{\underline{K}}\right]_{\mathrm{tr}} = \begin{bmatrix} -\sin\psi & -\cos\psi\cos\theta_{\mathrm{tr}} & -\sin\psi & \cos\psi\cos\theta_{\mathrm{tr}} \\ \cos\psi & -\sin\psi\cos\theta_{\mathrm{tr}} & \cos\psi & \sin\psi\cos\theta_{\mathrm{tr}} \\ -\left(\dfrac{n_2}{\eta_0}\right)\cos\psi\cos\theta_{\mathrm{tr}} & \left(\dfrac{n_2}{\eta_0}\right)\sin\psi & \left(\dfrac{n_2}{\eta_0}\right)\cos\psi\cos\theta_{\mathrm{tr}} & \left(\dfrac{n_2}{\eta_0}\right)\sin\psi \\ -\left(\dfrac{n_2}{\eta_0}\right)\sin\psi\cos\theta_{\mathrm{tr}} & -\left(\dfrac{n_2}{\eta_0}\right)\cos\psi & \left(\dfrac{n_2}{\eta_0}\right)\sin\psi\cos\theta_{\mathrm{tr}} & -\left(\dfrac{n_2}{\eta_0}\right)\cos\psi \end{bmatrix} . \tag{3.17}$$

3.3.1   STANDARD BOUNDARY CONDITIONS

Usually the tangential components of the auxiliary phasors are required to be continuous across interfaces [13]. That is, the following boundary conditions are imposed:

$$\left.\begin{aligned} \left[\underline{f}(0-)\right] &= \left[\underline{f}(0+)\right] \\ \left[\underline{f}(d_{\mathcal{A}}-)\right] &= \left[\underline{f}(d_{\mathcal{A}}+)\right] \\ \left[\underline{f}(\{d_{\mathcal{A}} + d_{\mathcal{B}}\}-)\right] &= \left[\underline{f}(\{d_{\mathcal{A}} + d_{\mathcal{B}}\}+)\right] \\ \left[\underline{f}(d-)\right] &= \left[\underline{f}(d+)\right] \end{aligned}\right\} . \tag{3.18}$$

Therefore, (3.12), (3.14), and (3.15) combine to yield the following linear relationship between the linear amplitudes of the transmitted plane wave and those of the incident and reflected plane waves:

$$
\begin{bmatrix} t_s \\ t_p \\ 0 \\ 0 \end{bmatrix} = \left[\underline{\underline{K}}\right]_{tr}^{-1} \cdot \left[\underline{\underline{M}}\right]^{\mathcal{C}} \cdot \left[\underline{\underline{M}}\right]^{\mathcal{B}} \cdot \left[\underline{\underline{M}}\right]^{\mathcal{A}} \cdot \left[\underline{\underline{K}}\right]_{inc} \cdot \begin{bmatrix} a_s \\ a_p \\ r_s \\ r_p \end{bmatrix}. \tag{3.19}
$$

The order of appearance of the transfer matrixes $\left[\underline{\underline{M}}\right]^{\mathcal{A}}$, $\left[\underline{\underline{M}}\right]^{\mathcal{B}}$, and $\left[\underline{\underline{M}}\right]^{\mathcal{C}}$ must not be disturbed in (3.19) because that order is governed by the sequential arrangement of the three layers. Even if all layers are electrically thin, a change in the order of appearance of the transfer matrixes amounts to an approximation [15] that may be acceptable only in certain conditions [16].

3.3.2   JUMP CONDITIONS

The TMM can also accommodate boundary conditions wherein there are discontinuities in the tangential components of the auxiliary phasors across layer interfaces, i.e., more general boundary conditions than those represented in (3.18). For example, such a discontinuity could arise if single-layer graphene were interposed between the layers $\mathcal{A}$ and $\mathcal{B}$. The appropriate boundary condition at the interface $z = d_{\mathcal{A}}$ would then be a jump condition written as [14]

$$
\left[\underline{f}(d_{\mathcal{A}}+)\right] = \left[\underline{\underline{S}}\right]^{\mathcal{A}/\mathcal{B}} \cdot \left[\underline{f}(d_{\mathcal{A}}-)\right], \tag{3.20}
$$

wherein the 4×4 matrix

$$
\left[\underline{\underline{S}}\right]^{\mathcal{A}/\mathcal{B}} = \begin{bmatrix} 1 & 0 & 0 & 0 \\ 0 & 1 & 0 & 0 \\ 0 & \sigma_{gr} & 1 & 0 \\ -\sigma_{gr} & 0 & 0 & 1 \end{bmatrix}, \tag{3.21}
$$

with σ_{gr} being the surface conductivity of graphene [17]. Then, (3.12), (3.14), and (3.15) would combine to yield

$$
\begin{bmatrix} t_s \\ t_p \\ 0 \\ 0 \end{bmatrix} = \left[\underline{\underline{K}}\right]_{tr}^{-1} \cdot \left[\underline{\underline{M}}\right]^{\mathcal{C}} \cdot \left[\underline{\underline{M}}\right]^{\mathcal{B}} \cdot \left[\underline{\underline{S}}\right]^{\mathcal{A}/\mathcal{B}} \cdot \left[\underline{\underline{M}}\right]^{\mathcal{A}} \cdot \left[\underline{\underline{K}}\right]_{inc} \cdot \begin{bmatrix} a_s \\ a_p \\ r_s \\ r_p \end{bmatrix} \tag{3.22}
$$

instead of (3.19).

Discontinuities in the tangential components of the auxiliary phasors across layer interfaces that arise if one (or more) of materials $\mathcal{A}$, $\mathcal{B}$, and $\mathcal{C}$ is a topological insulator can be similarly dealt with [18]. In this case, the protected surface states of the topological insulator are modeled via a surface admittance [19, 20].

For example, suppose that material $\mathcal{B}$ were an isotropic-dielectric topological insulator whose protected surface states are characterized by the surface admittance $\gamma_{\text{TI}}^{\mathcal{B}}$. The appropriate boundary conditions at the interfaces $z = d_{\mathcal{A}}$ and $z = d_{\mathcal{A}} + d_{\mathcal{B}}$ would be then written as [14]

$$\left[\underline{f}(d_{\mathcal{A}}+)\right] = \left[\underline{\underline{T}}\right]^{\mathcal{B}} \cdot \left[\underline{f}(d_{\mathcal{A}}-)\right] \tag{3.23}$$

and

$$\left[\underline{f}(\{d_{\mathcal{A}} + d_{\mathcal{B}}\}+)\right] = \left(\left[\underline{\underline{T}}\right]^{\mathcal{B}}\right)^{-1} \cdot \left[\underline{f}(\{d_{\mathcal{A}} + d_{\mathcal{B}}\}-)\right], \tag{3.24}$$

respectively, wherein the 4×4 matrix

$$\left[\underline{\underline{T}}\right]^{\mathcal{B}} = \begin{bmatrix} 1 & 0 & 0 & 0 \\ 0 & 1 & 0 & 0 \\ -\gamma_{\text{TI}}^{\mathcal{B}} & 0 & 1 & 0 \\ 0 & -\gamma_{\text{TI}}^{\mathcal{B}} & 0 & 1 \end{bmatrix}. \tag{3.25}$$

Then, (3.12), (3.14), and (3.15) would combine to yield

$$\begin{bmatrix} t_{\text{s}} \\ t_{\text{p}} \\ 0 \\ 0 \end{bmatrix} = \left[\underline{\underline{K}}\right]_{\text{tr}}^{-1} \cdot \left[\underline{\underline{M}}\right]^{\mathcal{C}} \cdot \left(\left[\underline{\underline{T}}\right]^{\mathcal{B}}\right)^{-1} \cdot \left[\underline{\underline{M}}\right]^{\mathcal{B}} \cdot \left[\underline{\underline{T}}\right]^{\mathcal{B}} \cdot \left[\underline{\underline{M}}\right]^{\mathcal{A}} \cdot \left[\underline{\underline{K}}\right]_{\text{inc}} \cdot \begin{bmatrix} a_{\text{s}} \\ a_{\text{p}} \\ r_{\text{s}} \\ r_{\text{p}} \end{bmatrix} \tag{3.26}$$

instead of (3.19).

3.4   LINEAR REFLECTANCES AND TRANSMITTANCES

By algebraic manipulation of (3.19), the linear reflection amplitudes r_{s} and r_{p} and the linear transmission amplitudes t_{s} and t_{p} may expressed as combinations of the linear incidence amplitudes a_{s} and a_{p}, per

$$\left.\begin{aligned} \begin{bmatrix} r_{\text{s}} \\ r_{\text{p}} \end{bmatrix} &= \begin{bmatrix} r_{\text{ss}} & r_{\text{sp}} \\ r_{\text{ps}} & r_{\text{pp}} \end{bmatrix} \cdot \begin{bmatrix} a_{\text{s}} \\ a_{\text{p}} \end{bmatrix} \\ \begin{bmatrix} t_{\text{s}} \\ t_{\text{p}} \end{bmatrix} &= \begin{bmatrix} t_{\text{ss}} & t_{\text{sp}} \\ t_{\text{ps}} & t_{\text{pp}} \end{bmatrix} \cdot \begin{bmatrix} a_{\text{s}} \\ a_{\text{p}} \end{bmatrix} \end{aligned}\right\}. \tag{3.27}$$

Herein, the elements of the 2×2 matrixes are the linear reflection and transmission coefficients, co-polarized coefficients having identical subscripts and cross-polarized coefficients having non-identical subscripts. For example, the cross-polarized reflection coefficient r_{sp} is associated with a p-polarized incident wave and an s-polarized reflected wave. Linear reflectances are derived by squaring the magnitudes of the corresponding linear reflection coefficients. For example,

$$R_{sp} = |r_{sp}|^2 \qquad (3.28)$$

is the linear reflectance corresponding to the linear reflection coefficient r_{sp}. Linear transmittances are related to the squared magnitudes of the corresponding linear transmission coefficients in the manner exemplified by

$$T_{sp} = \frac{n_2}{n_1} \frac{\text{Re}\{\cos\theta_{tr}\}}{\cos\theta_{inc}} |t_{sp}|^2 . \qquad (3.29)$$

Herein, T_{sp} is the cross-polarized linear transmittance associated with a p-polarized incident wave and an s-polarized transmitted wave. The other linear reflectances are defined in a similar manner to (3.28), and other linear transmittances in a similar manner to (3.29).

If it is assumed that none of the materials of layers $\mathcal{A}$, $\mathcal{B}$, and $\mathcal{C}$ is an active material, then the inequalities

$$\left. \begin{array}{l} R_{ss} + R_{ps} + T_{ss} + T_{ps} \leq 1 \\ R_{pp} + R_{sp} + T_{pp} + T_{sp} \leq 1 \end{array} \right\} \qquad (3.30)$$

must hold in order to comply with the principle of conservation of energy. If all materials of layers $\mathcal{A}$, $\mathcal{B}$, and $\mathcal{C}$ are neither active nor dissipative, then the inequalities (3.30) become equalities.

Linear absorptances are defined as

$$\left. \begin{array}{l} A_s = 1 - \left(R_{ss} + R_{ps} + T_{ss} + T_{ps} \right) \\ A_p = 1 - \left(R_{pp} + R_{sp} + T_{pp} + T_{sp} \right) \end{array} \right\} . \qquad (3.31)$$

These are useful quantities in the analysis of surface waves guided by the interface(s) of dissimilar materials. For example, localized peaks in the linear absorptances when viewed as functions of θ_{inc}, with k_0 and ψ being fixed, suggest the possible excitation of surface waves. Furthermore, if a peak is independent of the material thicknesses, beyond certain thresholds, then that is strong evidence of surface-wave propagation, but confirmation requires correlation with a specific solution of the underlying canonical boundary-value problem [3, 21].

3.5 CIRCULAR REFLECTANCES AND TRANSMITTANCES

While the decomposition of an arbitrarily polarized incident plane wave into s- and p-polarized components, as represented by (3.4), is conventional, it is not always convenient. For example, if the materials involved are chiral [22, 23], then it may be more convenient to decompose into left- and right-circularly polarized components. In that case, (3.4) are replaced by

$$
\left.
\begin{aligned}
\underline{E}_{\mathrm{inc}}(\underline{r}) &= \left(a_{\mathrm{L}}\, \frac{i\underline{s} - \underline{p}_{\mathrm{inc}}}{\sqrt{2}} - a_{\mathrm{R}}\, \frac{i\underline{s} + \underline{p}_{\mathrm{inc}}}{\sqrt{2}} \right) \\
&\quad \times \exp\{i\,[q\,(x \cos\psi + y \sin\psi) + k_0 n_1 z \cos\theta_{\mathrm{inc}}]\} \\
\underline{H}_{\mathrm{inc}}(\underline{r}) &= -i\,\frac{n_1}{\eta_0} \left(a_{\mathrm{L}}\, \frac{i\underline{s} - \underline{p}_{\mathrm{inc}}}{\sqrt{2}} + a_{\mathrm{R}}\, \frac{i\underline{s} + \underline{p}_{\mathrm{inc}}}{\sqrt{2}} \right) \\
&\quad \times \exp\{i\,[q\,(x \cos\psi + y \sin\psi) + k_0 n_1 z \cos\theta_{\mathrm{inc}}]\}
\end{aligned}
\right\}, \quad z < 0. \quad (3.32)
$$

Herein, the quantities

$$
\left.
\begin{aligned}
a_{\mathrm{L}} &= -\frac{i a_{\mathrm{s}} + a_{\mathrm{p}}}{\sqrt{2}} \\
a_{\mathrm{R}} &= \frac{i a_{\mathrm{s}} - a_{\mathrm{p}}}{\sqrt{2}}
\end{aligned}
\right\} \quad (3.33)
$$

are the amplitudes of the left- and right-circularly polarized components, respectively, of the incident plane wave. Correspondingly, (3.7) and (3.8) are replaced by

$$
\left.
\begin{aligned}
\underline{E}_{\mathrm{ref}}(\underline{r}) &= \left(-r_{\mathrm{L}}\, \frac{i\underline{s} - \underline{p}_{\mathrm{ref}}}{\sqrt{2}} + r_{\mathrm{R}}\, \frac{i\underline{s} + \underline{p}_{\mathrm{ref}}}{\sqrt{2}} \right) \\
&\quad \times \exp\{i\,[q\,(x \cos\psi + y \sin\psi) - k_0 n_1 z \cos\theta_{\mathrm{inc}}]\} \\
\underline{H}_{\mathrm{ref}}(\underline{r}) &= i\,\frac{n_1}{\eta_0} \left(r_{\mathrm{L}}\, \frac{i\underline{s} - \underline{p}_{\mathrm{ref}}}{\sqrt{2}} + r_{\mathrm{R}}\, \frac{i\underline{s} + \underline{p}_{\mathrm{ref}}}{\sqrt{2}} \right) \\
&\quad \times \exp\{i\,[q\,(x \cos\psi + y \sin\psi) - k_0 n_1 z \cos\theta_{\mathrm{inc}}]\}
\end{aligned}
\right\}, \quad z < 0, \quad (3.34)
$$

and

$$
\left.
\begin{aligned}
\underline{E}_{\mathrm{tr}}(\underline{r}) &= \left(t_{\mathrm{L}}\, \frac{i\underline{s} - \underline{p}_{\mathrm{tr}}}{\sqrt{2}} - t_{\mathrm{R}}\, \frac{i\underline{s} + \underline{p}_{\mathrm{tr}}}{\sqrt{2}} \right) \\
&\quad \times \exp\{i\,[q\,(x \cos\psi + y \sin\psi) + k_0 n_2 (z - d) \cos\theta_{\mathrm{tr}}]\} \\
\underline{H}_{\mathrm{tr}}(\underline{r}) &= -i\,\frac{n_2}{\eta_0} \left(t_{\mathrm{L}}\, \frac{i\underline{s} - \underline{p}_{\mathrm{tr}}}{\sqrt{2}} + t_{\mathrm{R}}\, \frac{i\underline{s} + \underline{p}_{\mathrm{tr}}}{\sqrt{2}} \right) \\
&\quad \times \exp\{i\,[q\,(x \cos\psi + y \sin\psi) + k_0 n_2 (z - d) \cos\theta_{\mathrm{tr}}]\}
\end{aligned}
\right\}, \quad z > d,
$$

$$
(3.35)
$$

respectively. Herein, the quantities

$$
\left.
\begin{aligned}
r_{\mathrm{L}} &= \frac{i r_{\mathrm{s}} + r_{\mathrm{p}}}{\sqrt{2}} \\
r_{\mathrm{R}} &= -\frac{i r_{\mathrm{s}} - r_{\mathrm{p}}}{\sqrt{2}}
\end{aligned}
\right\} \quad (3.36)
$$

are the amplitudes of the left- and right-circularly polarized components, respectively, of the reflected plane wave, while

$$
\left.\begin{aligned}
t_{\mathrm{L}} &= -\frac{i\,t_{\mathrm{s}} + t_{\mathrm{p}}}{\sqrt{2}} \\[2mm]
t_{\mathrm{R}} &= \frac{i\,t_{\mathrm{s}} - t_{\mathrm{p}}}{\sqrt{2}}
\end{aligned}\right\}
\tag{3.37}
$$

are the amplitudes of the left- and right-circularly polarized components, respectively, of the transmitted plane wave.

The circular reflection amplitudes r_{L} and r_{R}, and the circular transmission amplitudes t_{L} and t_{R} are related to the circular incidence amplitudes a_{L} and a_{R} by

$$
\left.\begin{aligned}
\begin{bmatrix} r_{\mathrm{L}} \\ r_{\mathrm{R}} \end{bmatrix} &= \begin{bmatrix} r_{\mathrm{LL}} & r_{\mathrm{LR}} \\ r_{\mathrm{RL}} & r_{\mathrm{RR}} \end{bmatrix} \cdot \begin{bmatrix} a_{\mathrm{L}} \\ a_{\mathrm{R}} \end{bmatrix} \\[3mm]
\begin{bmatrix} t_{\mathrm{L}} \\ t_{\mathrm{R}} \end{bmatrix} &= \begin{bmatrix} t_{\mathrm{LL}} & t_{\mathrm{LR}} \\ t_{\mathrm{RL}} & t_{\mathrm{RR}} \end{bmatrix} \cdot \begin{bmatrix} a_{\mathrm{L}} \\ a_{\mathrm{R}} \end{bmatrix}
\end{aligned}\right\},
\tag{3.38}
$$

wherein circular reflection and transmission coefficients are introduced as the elements of the 2×2 matrixes. As in the case of linear reflection and transmission coefficients, co-polarized circular reflection and transmission coefficients have both subscripts the same, whereas cross-polarized coefficients have two different subscripts. For example, the cross-polarized reflection coefficient r_{LR} is associated with a right-circularly polarized incident wave and a left-circularly polarized reflected wave.

The circular reflection and transmission coefficients can be expressed in terms of the linear reflection and transmission coefficients, and vice versa, as follows:

$$r_{ss} = -\frac{(r_{LL} + r_{RR}) - (r_{LR} + r_{RL})}{2}$$

$$r_{sp} = i \frac{(r_{LL} - r_{RR}) + (r_{LR} - r_{RL})}{2}$$

$$r_{ps} = -i \frac{(r_{LL} - r_{RR}) - (r_{LR} - r_{RL})}{2}$$

$$r_{pp} = -\frac{(r_{LL} + r_{RR}) + (r_{LR} + r_{RL})}{2}$$

$$t_{ss} = \frac{(t_{LL} + t_{RR}) - (t_{LR} + t_{RL})}{2}$$

$$t_{sp} = -i \frac{(t_{LL} - t_{RR}) + (t_{LR} - t_{RL})}{2}$$

$$t_{ps} = i \frac{(t_{LL} - t_{RR}) - (t_{LR} - t_{RL})}{2}$$

$$t_{pp} = \frac{(t_{LL} + t_{RR}) + (t_{LR} + t_{RL})}{2}$$

$$r_{LL} = -\frac{(r_{ss} + r_{pp}) + i (r_{sp} - r_{ps})}{2}$$

$$r_{LR} = \frac{(r_{ss} - r_{pp}) - i (r_{sp} + r_{ps})}{2}$$

$$r_{RL} = \frac{(r_{ss} - r_{pp}) + i (r_{sp} + r_{ps})}{2}$$

$$r_{RR} = -\frac{(r_{ss} + r_{pp}) - i (r_{sp} - r_{ps})}{2}$$

$$t_{LL} = \frac{(t_{ss} + t_{pp}) + i (t_{sp} - t_{ps})}{2}$$

$$t_{LR} = -\frac{(t_{ss} - t_{pp}) - i (t_{sp} + t_{ps})}{2}$$

$$t_{RL} = -\frac{(t_{ss} - t_{pp}) + i (t_{sp} + t_{ps})}{2}$$

$$t_{RR} = \frac{(t_{ss} + t_{pp}) - i (t_{sp} - t_{ps})}{2}$$

$$\tag{3.39}$$

Circular reflectances and transmittances are defined in an analogous manner to linear reflectances and transmittances. For example,

$$R_{LR} = |r_{LR}|^2 \tag{3.40}$$

is the cross-polarized circular reflectance corresponding to the circular reflectance coefficient r_{LR}, and

$$T_{LR} = \frac{n_2}{n_1} \frac{\text{Re}\{\cos\theta_{tr}\}}{\cos\theta_{inc}} |t_{LR}|^2 \tag{3.41}$$

is the cross-polarized circular transmittance corresponding to the circular transmittance coefficient t_{LR} (i.e., associated with a right-circularly polarized incident wave and a left-circularly polarized transmitted wave). The other circular reflectances are defined in a similar manner to (3.40), and other circular transmittances are defined in a similar manner to (3.41).

Analogously to the case for linear reflectances and transmittances, if it assumed that none of the three materials is active, then the inequalities

$$\left. \begin{array}{l} R_{LL} + R_{RL} + T_{LL} + T_{RL} \leq 1 \\ R_{RR} + R_{LR} + T_{RR} + T_{LR} \leq 1 \end{array} \right\} \tag{3.42}$$

must hold in order to comply with the principle of conservation of energy. If the materials of layers $\mathcal{A}$, $\mathcal{B}$, and $\mathcal{C}$ are neither active nor dissipative, then the inequalities (3.42) become equalities.

Circular absorptances are defined as

$$A_L = 1 - (R_{LL} + R_{RL} + T_{LL} + T_{RL}) \left.\begin{array}{c}\\\end{array}\right\}$$
$$A_R = 1 - (R_{RR} + R_{LR} + T_{RR} + T_{LR}) \left.\begin{array}{c}\\\end{array}\right\} . \qquad (3.43)$$

Similarly to the linear absorptances, the circular absorptances are useful quantities in the analysis of surface waves guided by the interface(s) of dissimilar materials. For example, localized peaks in the circular absorptances when viewed as functions of θ_{inc}, with k_0 and ψ being fixed, suggest the possible excitation of surface waves. Furthermore, if a peak is independent of the material thicknesses, beyond certain thresholds, then that is strong evidence of surface-wave propagation, but confirmation requires correlation with a specific solution of the underlying canonical boundary-value problem [3, 21].

3.6   SAMPLE NUMERICAL RESULTS

Plots of linear reflectances, transmittances, and absorptances vs. angles of incidence θ_{inc} and ψ are provided in this section in Fig. 3.2 for a representative multilayered slab comprising layers that are made of anisotropic and bianisotropic materials. Specifically, layer $\mathcal{A}$ is made of a biaxial bianisotropic material [1] specified by constitutive dyadics of the form

$$\underline{\underline{\sigma}}^{\mathcal{A}} = \sigma_{xx}^{\mathcal{A}} \, \underline{u}_x \underline{u}_x + \sigma_{yy}^{\mathcal{A}} \, \underline{u}_y \underline{u}_y + \sigma_{zz}^{\mathcal{A}} \, \underline{u}_z \underline{u}_z, \qquad \sigma \in \{\varepsilon, \xi, \zeta, \mu\}, \qquad (3.44)$$

with $\underline{\underline{\xi}}^{\mathcal{A}} = -\underline{\underline{\zeta}}^{\mathcal{A}}$; layer $\mathcal{B}$ is made of a uniaxial dielectric material [1] specified by constitutive dyadics of the form

$$\left.\begin{array}{l} \underline{\underline{\varepsilon}}^{\mathcal{B}} = \varepsilon_{xx}^{\mathcal{B}} \, \underline{u}_x \underline{u}_x + \varepsilon_{yy}^{\mathcal{B}} \, \underline{u}_y \underline{u}_y + \varepsilon_{zz}^{\mathcal{B}} \, \underline{u}_z \underline{u}_z \\[2mm] \underline{\underline{\xi}}^{\mathcal{B}} = \underline{\underline{0}} \\[2mm] \underline{\underline{\zeta}}^{\mathcal{B}} = \underline{\underline{0}} \\[2mm] \underline{\underline{\mu}}^{\mathcal{B}} = \mu_0 \underline{\underline{I}} \end{array}\right\} \qquad (3.45)$$

with $\varepsilon_{xx}^{\mathcal{B}} = \varepsilon_{zz}^{\mathcal{B}}$; and layer $\mathcal{C}$ is made of a Faraday chiral material [24] specified by constitutive dyadics of the form

$$\underline{\underline{\sigma}}^{\mathcal{C}} = \sigma_{xx}^{\mathcal{C}} \, \underline{u}_x \underline{u}_x + \sigma_{yy}^{\mathcal{C}} \, \underline{u}_y \underline{u}_y + \sigma_{zz}^{\mathcal{C}} \, \underline{u}_z \underline{u}_z + \sigma_{xy}^{\mathcal{C}} \underline{u}_x \underline{u}_y + \sigma_{yx}^{\mathcal{C}} \underline{u}_y \underline{u}_x, \qquad \sigma \in \{\varepsilon, \xi, \zeta, \mu\}, \quad (3.46)$$

with $\sigma_{xx}^{\mathcal{C}} = \sigma_{yy}^{\mathcal{C}}$, $\sigma_{xy}^{\mathcal{C}} = -\sigma_{yx}^{\mathcal{C}}$, and $\underline{\underline{\xi}}^{\mathcal{C}} = -\underline{\underline{\zeta}}^{\mathcal{C}}$. The parameters used for the calculations are specified in the caption to Fig. 3.2. The Mathematica™ code used to generate the numerical data plotted in Fig. 3.2 is provided in Appendix B.

For both s- and p-polarized incident light, Fig. 3.2 reveals that the linear absorptances are relatively large and remain almost constant as θ_{inc} increases from 0°, but as θ_{inc} approaches 90°

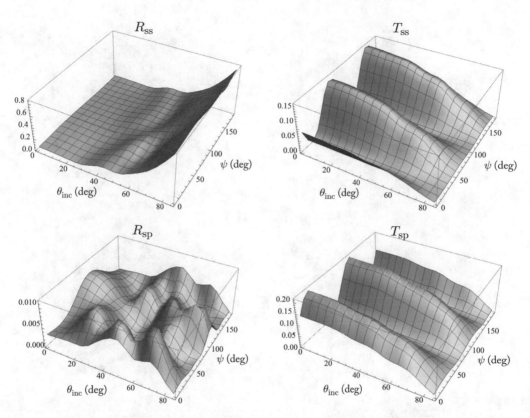

Figure 3.2: Linear reflectances, transmittances, and absorptances plotted against angles of incidence θ_{inc} and ψ for a bianisotropic multilayered slab with planar interfaces. Constitutive parameter values: $\varepsilon_{xx}^{\mathcal{A}} = (2.9 + 0.04i)\,\varepsilon_0$, $\varepsilon_{yy}^{\mathcal{A}} = (2.5 + 0.03i)\,\varepsilon_0$, $\varepsilon_{zz}^{\mathcal{A}} = (2.1 + 0.02i)\,\varepsilon_0$, $\xi_{xx}^{\mathcal{A}} = (0.1 + 0.0008i)\,i\sqrt{\varepsilon_0\mu_0}$, $\xi_{yy}^{\mathcal{A}} = (0.07 + 0.0006i)\,i\sqrt{\varepsilon_0\mu_0}$, $\xi_{zz}^{\mathcal{A}} = (0.06 + 0.0005i)\,i\sqrt{\varepsilon_0\mu_0}$, $\mu_{xx}^{\mathcal{A}} = (1.1 + 0.025i)\,\mu_0$, $\mu_{yy}^{\mathcal{A}} = (1.05 + 0.015i)\,\mu_0$, $\mu_{zz}^{\mathcal{A}} = (1.02 + 0.012i)\,\mu_0$; $\varepsilon_{xx}^{\mathcal{B}} = (2.4 + 0.03i)\,\varepsilon_0$, $\varepsilon_{yy}^{\mathcal{B}} = (1.9 + 0.02i)\,\varepsilon_0$; $\varepsilon_{xx}^{\mathcal{C}} = (4.1 + 0.08i)\,\varepsilon_0$, $\varepsilon_{zz}^{\mathcal{C}} = (3.4 + 0.06i)\,\varepsilon_0$, $\varepsilon_{xy}^{\mathcal{C}} = -(0.8 + 0.005i)\,i\varepsilon_0$, $\xi_{xx}^{\mathcal{C}} = (0.06 + 0.0012i)\,i\sqrt{\varepsilon_0\mu_0}$, $\xi_{zz}^{\mathcal{C}} = (0.04 + 0.001i)\,i\sqrt{\varepsilon_0\mu_0}$, $\xi_{xy}^{\mathcal{C}} = (0.01 + 0.0003i)\,\sqrt{\varepsilon_0\mu_0}$, $\mu_{xx}^{\mathcal{C}} = (1.3 + 0.04i)\,\mu_0$, $\mu_{zz}^{\mathcal{C}} = (1.1 + 0.03i)\,\mu_0$, and $\mu_{xy}^{\mathcal{C}} = -(0.35 + 0.003i)\,i\mu_0$. Also, $d_{\mathcal{A}} = 0.9\lambda_0$, $d_{\mathcal{B}} = 1.8\lambda_0$, $d_{\mathcal{C}} = 1.2\lambda_0$, with $\lambda_0 = 560$ nm. (*Continues.*)

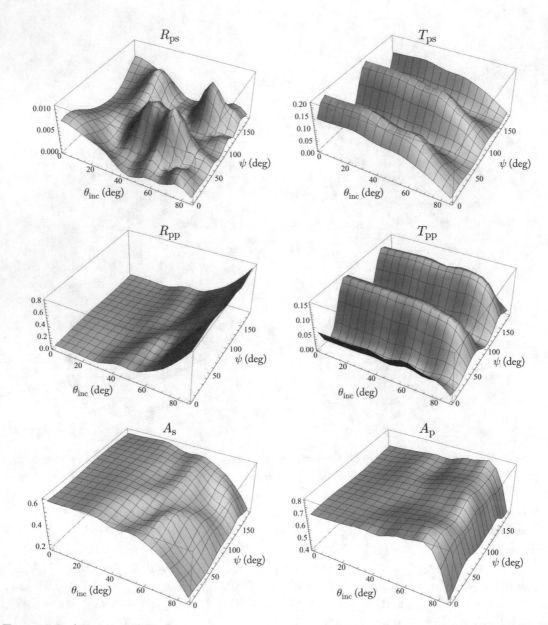

Figure 3.2: (*Continued.*) Reflectances, transmittances, and absorptances calculated for a bian-isotropic multilayered slab.

both absorptances approach zero. Also, both linear absorptances vary little as ψ varies. The cross-polarized reflectances and transmittances likewise generally approach zero as θ_{inc} approaches 90°, and these quantities vary substantially as ψ varies. Whether the incident light is s- or p-polarized, the co-polarized reflectances approach unity and the co-polarized transmittances approach zero as θ_{inc} approaches 90°; the co-polarized reflectances are largely insensitive to variations in ψ, whereas the co-polarized transmittances oscillate markedly as ψ increases.

3.7 REFERENCES

[1] Mackay, T. G. and Lakhtakia, A. 2019. *Electromagnetic Anisotropy and Bianisotropy: A Field Guide*, 2nd ed. (Singapore: World Scientific). DOI: 10.1142/11351. 33, 44

[2] Boardman, A. D., Ed., 1982. *Electromagnetic Surface Modes* (Chicester, UK, Wiley). 33

[3] Polo, J. A., Jr., Mackay, T. G., and Lakhtakia, A. 2013. *Electromagnetic Surface Waves: A Modern Perspective* (Waltham, MA, Elsevier). 33, 40, 44

[4] Turbadar, T. 1959. Complete absorption of light by thin metal films, *Proc. Phys. Soc.*, 73:40–44. DOI: 10.1088/0370-1328/73/1/307. 33

[5] Turbadar, T. 1964. Complete absorption of plane polarized light by thin metal films, *Opt. Acta*, 11:207–210. DOI: 10.1080/713817875. 33

[6] Kretschmann, E. and Raether, H. 1968. Radiative decay of nonradiative surface plasmons excited by light, *Z. Naturforsch. A*, 23:2135–2136. DOI: 10.1515/zna-1968-1247. 33

[7] Otto, A. 1968. Excitation of nonradiative surface plasma waves in silver by the method of frustrated total reflection, *Z. Phys.*, 216:398–410. DOI: 10.1007/bf01391532. 33

[8] Lakhtakia, A. and Messier, R. 2005. *Sculptured Thin Films: Nanoengineered Morphology and Optics* (Bellingham, WA, SPIE Press). DOI: 10.1117/3.585322. 33

[9] Yeh, P., Yariv, A., and Hong, C.-S. 1977. Electromagnetic propagation in periodic stratified media, I. General theory, *J. Opt. Soc. Am.*, 67:423–438. DOI: 10.1364/josa.67.000423. 33

[10] Martorell, J., Sprung, D. W. L., and Morozov, G. V. 2006. Surface TE waves on 1D photonic crystals, *J. Opt. A: Pure Appl. Opt.*, 8:630–638. DOI: 10.1088/1464-4258/8/8/003. 33

[11] Lakhtakia, A. and Polo, J. A., Jr. 2007. Dyakonov–Tamm wave at the planar interface of a chiral sculptured thin film and an isotropic dielectric material, *J. Eur. Opt. Soc.—Rapid Pub.*, 2:07021. DOI: 10.2971/jeos.2007.07021. 33

[12] Pulsifer, D .P., Faryad, M., and Lakhtakia, A. 2013. Observation of the Dyakonov–Tamm wave, *Phys. Rev. Lett.*, 111:243902. DOI: 10.1103/physrevlett.111.243902. 33

[13] Chen, H. C. 1983. *Theory of Electromagnetic Waves* (New York, McGraw-Hill). 37

[14] Chiadini, F., Scaglione, A., Fiumara, V., Shuba, M. V., and Lakhtakia, A. 2019. Effect of chemical potential on Dyakonov–Tamm waves guided by a graphene-coated structurally chiral medium, *J. Opt.*, 21:055002, UK. DOI: 10.1088/2040-8986/ab137f.
Chiadini, F., Scaglione, A., Fiumara, V., Shuba, M. V., and Lakhtakia, A. 2019. Effect of chemical potential on Dyakonov–Tamm waves guided by a graphene-coated structurally chiral medium, *J. Opt.*, 21:079501, UK. (erratum) DOI: 10.1088/2040-8986/ab2a57. 38, 39

[15] Lakhtakia, A. 2003. Comment on 'Analytical solution of non-homogeneous anisotropic wave equations based on differential transfer matrices,' *J. Opt. A: Pure Appl. Opt.*, 5:432–433. DOI: 10.1088/1464-4258/5/4/401. 38

[16] Lakhtakia, A. and Krowne, C. M. 2003. Restricted equivalence of paired epsilon-negative and mu-negative layers to a negative phase-velocity material (*alias* left-handed material), *Optik*, 114:305–307. DOI: 10.1078/0030-4026-00266. 38

[17] Depine, R. A. 2016. *Graphene Optics: Electromagnetic Solution of Canonical Problems* (San Rafael, CA, Morgan & Claypool), IOP Concise Physics. DOI: 10.1088/978-1-6817-4309-7. 38

[18] Hasan, M. Z. and Kane, C. L. 2010. Topological insulators, *Rev. Mod. Phys.*, 82:3045–3067. DOI: 10.1103/RevModPhys.82.3045. 39

[19] Lakhtakia, A. and Mackay, T. G. 2016. Classical electromagnetic model of surface states in topological insulators, *J. Nanophoton.*, 10:033004. DOI: 10.1117/1.jnp.10.033004. 39

[20] Diovisalvi, A., Lakhtakia, A., Fiumara, V., and Chiadini, F. 2017. Bilaterally asymmetric reflection and transmission of light by a grating structure containing a topological insulator, *Opt. Commun.*, 398:67–76. DOI: 10.1016/j.optcom.2017.04.017. 39

[21] Mackay, T. G. 2019. On the identification of surface waves in numerical studies, *Plasmonics*, 14:1–2. DOI: 10.1007/s11468-018-0770-y. 40, 44

[22] Lakhtakia, A. 1994. *Beltrami Fields in Chiral Media* (Singapore, World Scientific). DOI: 10.1142/2031. 40

[23] Mackay, T. G. and Lakhtakia, A. 2010. Negatively refracting chiral metamaterials: A review, *SPIE Rev.*, 1:018003. DOI: 10.1117/6.0000003. 40

[24] Weiglhofer, W. S. and Lakhtakia, A. 1998. The correct constitutive relations of chiroplasmas and chiroferrites, *Microw. Opt. Technol. Lett.*, 17:405–408. DOI: 10.1002/(sici)1098-2760(19980420)17:6%3C405::aid-mop17%3E3.0.co;2-z. 44

CHAPTER 4

Bianisotropic Slab with Periodically Corrugated Interfaces

As in Chapter 3, the TMM is developed in this chapter for the planewave-response characteristics of a multilayered slab comprising three layers of materials labeled $\mathcal{A}$, $\mathcal{B}$, and $\mathcal{C}$. Each layer is made of a different homogeneous bianisotropic material [1]. However, unlike Chapter 3, the interface of layers $\mathcal{A}$ and $\mathcal{B}$ is periodically corrugated while the interface of layers $\mathcal{B}$ and $\mathcal{C}$ is planar, as is schematically illustrated in Fig. 4.1. This arrangement is appropriate for the analysis of surface waves excited in grating-coupled configurations [2, 3]. It is also applicable to reflection-transmission studies involving a multilayered slab with one or more constituent layers made of bianisotropic (or anisotropic) materials that are nonhomogeneous in the thickness direction; the nonhomogeneity may be accommodated by a piecewise-uniform approximation [3, 4]. Such applications arise when certain sculptured thin films [4] are incorporated in the multilayered slab as well as in the analysis of Dyakonov–Tamm surface waves [5, 6].

4.1 PRELIMINARIES

As in Chapter 3, consider a multilayered slab comprising three constituent layers, labeled $\mathcal{A}$, $\mathcal{B}$, and $\mathcal{C}$, each of infinite extent in the xy plane. The plane $z = 0$ forms one face of layer $\mathcal{A}$. The interface of layers $\mathcal{A}$ and $\mathcal{B}$ is $z = d_{\mathcal{A}} + g(x, y)$, with $g(x, y)$ being the grating function which is periodic in the x and y directions, in general. The interface of layers $\mathcal{B}$ and $\mathcal{C}$ is the plane $z = d_{\mathcal{A}} + d_{\mathcal{B}}$. Finally, the other face of layer $\mathcal{C}$ is the plane $z = d$ with $d = d_{\mathcal{A}} + d_{\mathcal{B}} + d_{\mathcal{C}}$. Thus, the geometry considered is this chapter is almost the same as that in Chapter 3, the only difference being that the planar interface $z = d_{\mathcal{A}}$ in Chapter 3 is here replaced by the periodically corrugated interface $z = d_{\mathcal{A}} + g(x, y)$.

The periodically corrugated interface $z = d_{\mathcal{A}} + g(x, y)$ can be singly periodic, i.e.,

$$g(x, y) \equiv g(x) = g(x \pm L_x), \tag{4.1}$$

where L_x is the period along the x axis, as is schematically illustrated in Fig. 4.1. But, more generally, this interface is doubly periodic, i.e.,

$$g(x, y) = g(x \pm L_x, y) = g(x, y \pm L_y), \tag{4.2}$$

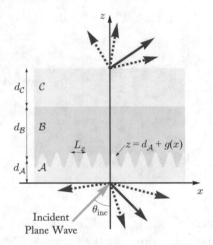

Figure 4.1: Schematic representation of the reflection and transmission of a plane wave by a multilayered slab comprising layers $\mathcal{A}$, $\mathcal{B}$, and $\mathcal{C}$. The layers $\mathcal{A}$ and $\mathcal{B}$ have a periodically corrugated interface $z = d_\mathcal{A} + g(x)$, and the layers $\mathcal{B}$ and $\mathcal{C}$ have a planar interface $z = d_\mathcal{A} + d_\mathcal{B}$. The direction of propagation of the incident plane wave is restricted in this figure to the xz plane, but the theory is not restricted. The angle of incidence $\theta_{\mathrm{inc}} \in [0, \pi/2)$.

where L_x and L_y are the periods along the x and y axes, respectively. The constraints

$$\left.\begin{array}{l} g(x, y) < d_\mathcal{B} \\ g(x, y) > -d_\mathcal{A} \end{array}\right\} , \quad \forall x \in (-\infty, \infty), \quad \forall y \in (-\infty, \infty) , \tag{4.3}$$

must be satisfied by the grating function for the formulation presented in this chapter, but a modified version can be used if otherwise. Let $g_{\max} > 0$ denote the maximum value of $g(x, y)$ and $g_{\min} < 0$ denote the minimum value of $g(x, y)$. Then the trough-to-crest height of the interface is $L_g = g_{\max} - g_{\min}$. For the grating-coupled excitation of surface waves, L_g can be much smaller than the free-space wavelength λ_0 [3].

The grating function can be sinusoidal [7], as illustrated in Fig. 4.1, but it need not be [8–10]. Indeed, the formulation presented applies to any periodic corrugation satisfying the constraints (4.3) and is independent of the particular geometry prescribed by $g(x, y)$ within each period. Parenthetically, compound gratings which comprise several periods each of two or more simple gratings can be used to excite multiple surface waves over a range of angles of incidence and wavelengths [11, 12].

Each of the layers $\mathcal{A}$, $\mathcal{B}$, and $\mathcal{C}$ is made of a different homogeneous bianisotropic material. Thus, the constitutive relations of the three materials are

$$\left.\begin{array}{l} \underline{D}(\underline{r}) = \underline{\underline{\varepsilon}}^\mathcal{A} \cdot \underline{E}(\underline{r}) + \underline{\underline{\xi}}^\mathcal{A} \cdot \underline{H}(\underline{r}) \\ \underline{B}(\underline{r}) = \underline{\underline{\zeta}}^\mathcal{A} \cdot \underline{E}(\underline{r}) + \underline{\underline{\mu}}^\mathcal{A} \cdot \underline{H}(\underline{r}) \end{array}\right\} , \quad 0 < z < d_\mathcal{A} + g(x, y) , \tag{4.4}$$

$$\left.\begin{array}{l} \underline{D}(\underline{r}) = \underline{\underline{\varepsilon}}^{\mathcal{B}} \cdot \underline{E}(\underline{r}) + \underline{\underline{\xi}}^{\mathcal{B}} \cdot \underline{H}(\underline{r}) \\ \underline{B}(\underline{r}) = \underline{\underline{\zeta}}^{\mathcal{B}} \cdot \underline{E}(\underline{r}) + \underline{\underline{\mu}}^{\mathcal{B}} \cdot \underline{H}(\underline{r}) \end{array}\right\}, \quad d_{\mathcal{A}} + g(x,y) < z < d_{\mathcal{A}} + d_{\mathcal{B}}, \qquad (4.5)$$

and

$$\left.\begin{array}{l} \underline{D}(\underline{r}) = \underline{\underline{\varepsilon}}^{\mathcal{C}} \cdot \underline{E}(\underline{r}) + \underline{\underline{\xi}}^{\mathcal{C}} \cdot \underline{H}(\underline{r}) \\ \underline{B}(\underline{r}) = \underline{\underline{\zeta}}^{\mathcal{C}} \cdot \underline{E}(\underline{r}) + \underline{\underline{\mu}}^{\mathcal{C}} \cdot \underline{H}(\underline{r}) \end{array}\right\}, \quad d_{\mathcal{A}} + d_{\mathcal{B}} < z < d. \qquad (4.6)$$

All constitutive dyadics in (4.4)–(4.6) have the general form given in (2.29) and (2.30), with the symbols ε, ξ, ζ, and μ therein replaced by, respectively, $\varepsilon^{\gamma}, \xi^{\gamma}, \zeta^{\gamma}$, and μ^{γ}, $\gamma \in \{\mathcal{A}, \mathcal{B}, \mathcal{C}\}$.

A homogeneous isotropic dielectric material fills the half-space $z < 0$. Its refractive index n_1 is taken to be real valued and positive. Likewise, the half-space $z > d$ is filled with a homogeneous isotropic dielectric material characterized by the refractive index n_2 which is real valued and positive.

4.2   INCIDENT PLANE WAVE

Suppose that a plane wave propagates in the half-space $z < 0$ at an angle $\theta_{\text{inc}} \in [0°, 90°)$ with respect to the z axis and at an angle $\psi \in [0°, 360°)$ with respect to the x axis in the xy plane. This plane wave is incident on the plane $z = 0$. The polarization state of this incident plane wave is arbitrary, and the associated field phasors can be represented in terms of s- and p- polarized components, analogously to (3.4). Due to the periodically corrugated interface of layers $\mathcal{A}$ and $\mathcal{B}$, those field phasors are best expressed as the sum of linear Floquet harmonics [3, 13] as

$$\left.\begin{array}{l} \underline{E}_{\text{inc}}(\underline{r}) = \displaystyle\sum_{m\in\mathbb{Z}} \sum_{n\in\mathbb{Z}} \left\{ \left(a_{\text{s}}^{(m,n)} \underline{s}^{(m,n)} + a_{\text{p}}^{(m,n)} \underline{p}_{\text{inc}}^{(m,n)} \right) \right. \\ \qquad\qquad \left. \times \exp\left[i \left(\underline{\kappa}^{(m,n)} + \alpha_1^{(m,n)} \underline{u}_{\text{z}} \right) \cdot \underline{r} \right] \right\} \\[4mm] \underline{H}_{\text{inc}}(\underline{r}) = n_1 \eta_0^{-1} \displaystyle\sum_{m\in\mathbb{Z}} \sum_{n\in\mathbb{Z}} \left\{ \left(a_{\text{s}}^{(m,n)} \underline{p}_{\text{inc}}^{(m,n)} - a_{\text{p}}^{(m,n)} \underline{s}^{(m,n)} \right) \right. \\ \qquad\qquad \left. \times \exp\left[i \left(\underline{\kappa}^{(m,n)} + \alpha_1^{(m,n)} \underline{u}_{\text{z}} \right) \cdot \underline{r} \right] \right\} \end{array}\right\}, \quad z < 0. \quad (4.7)$$

Herein, the quantities

$$
\left.\begin{aligned}
\underline{\kappa}^{(m,n)} &= k_x^{(m)} \underline{u}_x + k_y^{(n)} \underline{u}_y \\
k_x^{(m)} &= k_0 n_1 \sin \theta_{\text{inc}} \cos \psi + m(2\pi/L_x) \\
k_y^{(n)} &= k_0 n_1 \sin \theta_{\text{inc}} \sin \psi + n(2\pi/L_y) \\
k_{xy}^{(m,n)} &= +\sqrt{\underline{\kappa}^{(m,n)} \cdot \underline{\kappa}^{(m,n)}} \\
\alpha_1^{(m,n)} &= +\sqrt{(k_0 n_1)^2 - \underline{\kappa}^{(m,n)} \cdot \underline{\kappa}^{(m,n)}} \\
\underline{s}^{(m,n)} &= -\frac{k_y^{(n)}}{k_{xy}^{(m,n)}} \underline{u}_x + \frac{k_x^{(m)}}{k_{xy}^{(m,n)}} \underline{u}_y \\
\underline{p}_{\text{inc}}^{(m,n)} &= -\left(\frac{k_x^{(m)}}{k_{xy}^{(m,n)}} \underline{u}_x + \frac{k_y^{(n)}}{k_{xy}^{(m,n)}} \underline{u}_y \right) \frac{\alpha_1^{(m,n)}}{k_0 n_1} + \frac{k_{xy}^{(m,n)}}{k_0 n_1} \underline{u}_z
\end{aligned} \right\}, \tag{4.8}
$$

while the amplitudes

$$
\left.\begin{aligned}
a_s^{(m,n)} &= a_s^{(0,0)} \delta_{m0} \delta_{n0} \\
a_p^{(m,n)} &= a_p^{(0,0)} \delta_{m0} \delta_{n0}
\end{aligned} \right\} \tag{4.9}
$$

are defined using the Kronecker delta

$$
\delta_{nn'} = \begin{cases} 1, & n = n' \\ 0, & n \neq n' \end{cases} . \tag{4.10}
$$

Each Floquet harmonic is identified by an index pair (m,n), with each index belonging to the set $\mathbb{Z} = \{0, \pm1, \pm2, \ldots\}$. The relations (4.9) ensure that only the Floquet harmonic with index pair $(0,0)$ contributes to the incident plane wave.

The scalar quantity $\alpha_1^{(m,n)}$ is positive real if $k_0 n_1 > k_{xy}^{(m,n)}$, in which case the corresponding Floquet harmonic in (4.7) is a plane wave that can transport energy as $z \to \infty$, provided that all space is filled with the dielectric material of refractive index n_1. On the other hand, $\alpha_1^{(m,n)}$ is positive imaginary if $k_0 n_1 < k_{xy}^{(m,n)}$, in which case the corresponding Floquet harmonic in (4.7) is evanescent as $z \to \infty$.

4.3   REFLECTED AND TRANSMITTED FIELD PHASORS

Due to the periodic corrugation of the interface of layers $\mathcal{A}$ and $\mathcal{B}$, the representations (3.7) and (3.8) of the reflected and transmitted field phasors are inadequate. Instead these phasors must also be expressed as sums of linear Floquet harmonics [3, 13]. Thus, the reflected field phasors are written as

$$
\left.
\begin{aligned}
\underline{E}_{\mathrm{ref}}(\underline{r}) &= \sum_{m\in\mathbb{Z}}\sum_{n\in\mathbb{Z}}\left\{\left(r_{\mathrm{s}}^{(m,n)}\,\underline{s}^{(m,n)} + r_{\mathrm{p}}^{(m,n)}\,\underline{p}_{\mathrm{ref}}^{(m,n)}\right)\right. \\
&\qquad \left. \times \exp\left[i\left(\underline{\kappa}^{(m,n)} - \alpha_1^{(m,n)}\underline{u}_{\mathrm{z}}\right)\cdot\underline{r}\right]\right\} \\
\underline{H}_{\mathrm{ref}}(\underline{r}) &= n_1\eta_0^{-1}\sum_{m\in\mathbb{Z}}\sum_{n\in\mathbb{Z}}\left\{\left(r_{\mathrm{s}}^{(m,n)}\,\underline{p}_{\mathrm{ref}}^{(m,n)} - r_{\mathrm{p}}^{(m,n)}\,\underline{s}^{(m,n)}\right)\right. \\
&\qquad \left. \times \exp\left[i\left(\underline{\kappa}^{(m,n)} - \alpha_1^{(m,n)}\underline{u}_{\mathrm{z}}\right)\cdot\underline{r}\right]\right\}
\end{aligned}
\right\}, \quad z < 0, \quad (4.11)
$$

where

$$
\underline{p}_{\mathrm{ref}}^{(m,n)} = \left(\frac{k_x^{(m)}}{k_{xy}^{(m,n)}}\underline{u}_{\mathrm{x}} + \frac{k_y^{(n)}}{k_{xy}^{(m,n)}}\underline{u}_{\mathrm{y}}\right)\frac{\alpha_1^{(m,n)}}{k_0 n_1} + \frac{k_{xy}^{(m,n)}}{k_0 n_1}\underline{u}_{\mathrm{z}}. \tag{4.12}
$$

The reflection amplitudes $r_{\mathrm{s}}^{(m,n)}$ and $r_{\mathrm{p}}^{(m,n)}$ have to be determined for all $m \in \mathbb{Z}$ and $n \in \mathbb{Z}$. If $\alpha_1^{(m,n)} > 0$, then the corresponding Floquet harmonic in (4.11) is a plane wave that transports energy as $z \to -\infty$. If $-i\alpha_1^{(m,n)} > 0$, then the Floquet harmonic is evanescent as $z \to -\infty$.

In a similar vein, the transmitted field phasors are written as

$$
\left.
\begin{aligned}
\underline{E}_{\mathrm{tr}}(\underline{r}) &= \sum_{m\in\mathbb{Z}}\sum_{n\in\mathbb{Z}}\left\{\left(t_{\mathrm{s}}^{(m,n)}\,\underline{s}^{(m,n)} + t_{\mathrm{p}}^{(m,n)}\,\underline{p}_{\mathrm{tr}}^{(m,n)}\right)\right. \\
&\qquad \left. \times \exp\left[i\left(\underline{\kappa}^{(m,n)} + \alpha_2^{(m,n)}\underline{u}_{\mathrm{z}}\right)\cdot\left(\underline{r} - d\underline{u}_{\mathrm{z}}\right)\right]\right\} \\
\underline{H}_{\mathrm{tr}}(\underline{r}) &= n_2\eta_0^{-1}\sum_{m\in\mathbb{Z}}\sum_{n\in\mathbb{Z}}\left\{\left(t_{\mathrm{s}}^{(m,n)}\,\underline{p}_{\mathrm{tr}}^{(m,n)} - t_{\mathrm{p}}^{(m,n)}\,\underline{s}^{(m,n)}\right)\right. \\
&\qquad \left. \times \exp\left[i\left(\underline{\kappa}^{(m,n)} + \alpha_2^{(m,n)}\underline{u}_{\mathrm{z}}\right)\cdot\left(\underline{r} - d\underline{u}_{\mathrm{z}}\right)\right]\right\}
\end{aligned}
\right\}, \quad z > d, \quad (4.13)
$$

wherein

$$
\left.
\begin{aligned}
\alpha_2^{(m,n)} &= +\sqrt{(k_0 n_2)^2 - \underline{\kappa}^{(m,n)}\cdot\underline{\kappa}^{(m,n)}} \\
\underline{p}_{\mathrm{tr}}^{(m,n)} &= -\left(\frac{k_x^{(m)}}{k_{xy}^{(m,n)}}\underline{u}_{\mathrm{x}} + \frac{k_y^{(n)}}{k_{xy}^{(m,n)}}\underline{u}_{\mathrm{y}}\right)\frac{\alpha_2^{(m,n)}}{k_0 n_2} + \frac{k_{xy}^{(m,n)}}{k_0 n_2}\underline{u}_{\mathrm{z}}
\end{aligned}
\right\}. \tag{4.14}
$$

The transmission amplitudes $t_{\mathrm{s}}^{(m,n)}$ and $t_{\mathrm{p}}^{(m,n)}$ have to be determined for all $m \in \mathbb{Z}$ and $n \in \mathbb{Z}$. The quantity $\alpha_2^{(m,n)} > 0$ if $k_0 n_2 > k_{xy}^{(m,n)}$, in which case the corresponding Floquet harmonic in (4.13) is a plane wave that transports energy as $z \to \infty$. The quantity $-i\alpha_2^{(m,n)} > 0$ if $k_0 n_2 < k_{xy}^{(m,n)}$, in which case the corresponding Floquet harmonic is evanescent as $z \to \infty$.

The sums on the right sides of (4.11) and (4.13) contain *specular* terms, which are identified by the indexes $m = n = 0$. The remaining terms in these sums are *nonspecular* terms whose existence stems from the periodic nature of the function $g(x,y)$. If $g(x,y) \equiv g(x)$, we have $n \in \{0\}$; if $g(x,y) \equiv g(y)$, then $m \in \{0\}$. In the degenerate case $g(x,y) = 0$, there are no nonspecular terms and the formulation provided in this chapter simplifies to the one in Chapter 3. The reflection and transmission amplitudes are conveniently determined via the rigorous coupled-wave approach (RCWA) described in Section 4.6.

4.4   LINEAR REFLECTANCES AND TRANSMITTANCES

The reflection and transmission amplitudes of order (m, n) are related to the amplitudes of the incident plane wave by the relations

$$
\left.
\begin{array}{l}
\begin{bmatrix} r_{\mathrm{s}}^{(m,n)} \\ r_{\mathrm{p}}^{(m,n)} \end{bmatrix}
=
\begin{bmatrix} r_{\mathrm{ss}}^{(m,n)} & r_{\mathrm{sp}}^{(m,n)} \\ r_{\mathrm{ps}}^{(m,n)} & r_{\mathrm{pp}}^{(m,n)} \end{bmatrix}
\cdot
\begin{bmatrix} a_{\mathrm{s}}^{(0,0)} \\ a_{\mathrm{p}}^{(0,0)} \end{bmatrix}
\\[18pt]
\begin{bmatrix} t_{\mathrm{s}}^{(m,n)} \\ t_{\mathrm{p}}^{(m,n)} \end{bmatrix}
=
\begin{bmatrix} t_{\mathrm{ss}}^{(m,n)} & t_{\mathrm{sp}}^{(m,n)} \\ t_{\mathrm{ps}}^{(m,n)} & t_{\mathrm{pp}}^{(m,n)} \end{bmatrix}
\cdot
\begin{bmatrix} a_{\mathrm{s}}^{(0,0)} \\ a_{\mathrm{p}}^{(0,0)} \end{bmatrix}
\end{array}
\right\}, \tag{4.15}
$$

wherein the elements in the 2×2 matrixes are the reflection and transmission coefficients of order (m, n). The specular coefficients are those of order $(0, 0)$, whereas all other coefficients are nonspecular.

In an analogous manner to that presented in Section 3.4, the reflection coefficients of order (m, n) give rise to linear reflectances of order (m, n), as exemplified by

$$
R_{\mathrm{sp}}^{(m,n)} = \frac{\mathrm{Re}\left\{\alpha_1^{(m,n)}\right\}}{\alpha_1^{(0,0)}} |r_{\mathrm{sp}}^{(m,n)}|^2 . \tag{4.16}
$$

Likewise, the transmission coefficients of order (m, n) deliver linear transmittances of order (m, n); e.g.,

$$
T_{\mathrm{sp}}^{(m,n)} = \frac{\mathrm{Re}\left\{\alpha_2^{(m,n)}\right\}}{\alpha_1^{(0,0)}} |t_{\mathrm{sp}}^{(m,n)}|^2 . \tag{4.17}
$$

The specular linear reflectances and transmittances are those of order $(0, 0)$, while all other linear reflectances and transmittances are nonspecular.

Linear absorptances for s- and p-polarization states are defined as

$$
\left.
\begin{array}{l}
A_{\mathrm{s}} = 1 - \displaystyle\sum_{m\in\mathbb{Z}} \sum_{n\in\mathbb{Z}} \left(R_{\mathrm{ss}}^{(m,n)} + R_{\mathrm{ps}}^{(m,n)} + T_{\mathrm{ss}}^{(m,n)} + T_{\mathrm{ps}}^{(m,n)} \right) \\[20pt]
A_{\mathrm{p}} = 1 - \displaystyle\sum_{m\in\mathbb{Z}} \sum_{n\in\mathbb{Z}} \left(R_{\mathrm{pp}}^{(m,n)} + R_{\mathrm{sp}}^{(m,n)} + T_{\mathrm{pp}}^{(m,n)} + T_{\mathrm{sp}}^{(m,n)} \right)
\end{array}
\right\}, \tag{4.18}
$$

respectively. Provided that the materials of all three layers are passive, the linear absorptances are bounded above by unity and below by zero [1]. If all three materials are neither dissipative nor active, then $A_{\mathrm{s}} = A_{\mathrm{p}} = 0$.

4.5 CIRCULAR REFLECTANCES AND TRANSMITTANCES

Instead of the decompositions in terms of linear Floquet harmonics presented in Sections 4.2 and 4.3, the incident plane wave and the reflected and transmitted field phasors can be expressed in terms of circularly polarized Floquet harmonics, as follows. The incident plane wave is represented as

$$
\left.\begin{aligned}
\underline{E}_{\mathrm{inc}}(\underline{r}) &= \sum_{m\in\mathbb{Z}}\sum_{n\in\mathbb{Z}}\left\{\left(a_{\mathrm{L}}^{(m,n)}\frac{i\underline{s}^{(m,n)}-\underline{p}_{\mathrm{inc}}^{(m,n)}}{\sqrt{2}}-a_{\mathrm{R}}^{(m,n)}\frac{i\underline{s}^{(m,n)}+\underline{p}_{\mathrm{inc}}^{(m,n)}}{\sqrt{2}}\right)\right.\\
&\qquad\left.\times\exp\left[i\left(\underline{\kappa}^{(m,n)}+\alpha_1^{(m,n)}\underline{u}_{\mathrm{z}}\right)\boldsymbol{\cdot}\underline{r}\right]\right\}\\
\underline{H}_{\mathrm{inc}}(\underline{r}) &= -in_1\eta_0^{-1}\sum_{m\in\mathbb{Z}}\sum_{n\in\mathbb{Z}}\left\{\left(a_{\mathrm{L}}^{(m,n)}\frac{i\underline{s}^{(m,n)}-\underline{p}_{\mathrm{inc}}^{(m,n)}}{\sqrt{2}}+a_{\mathrm{R}}^{(m,n)}\frac{i\underline{s}^{(m,n)}+\underline{p}_{\mathrm{inc}}^{(m,n)}}{\sqrt{2}}\right)\right.\\
&\qquad\left.\times\exp\left[i\left(\underline{\kappa}^{(m,n)}+\alpha_1^{(m,n)}\underline{u}_{\mathrm{z}}\right)\boldsymbol{\cdot}\underline{r}\right]\right\}\\
& \qquad\qquad\qquad\qquad z<0,
\end{aligned}\right\},\qquad(4.19)
$$

while the reflected and transmitted field phasors are expressed as

$$
\left.\begin{aligned}
\underline{E}_{\mathrm{ref}}(\underline{r}) &= \sum_{m\in\mathbb{Z}}\sum_{n\in\mathbb{Z}}\left\{\left(-r_{\mathrm{L}}^{(m,n)}\frac{i\underline{s}^{(m,n)}-\underline{p}_{\mathrm{ref}}^{(m,n)}}{\sqrt{2}}+r_{\mathrm{R}}^{(m,n)}\frac{i\underline{s}^{(m,n)}+\underline{p}_{\mathrm{ref}}^{(m,n)}}{\sqrt{2}}\right)\right.\\
&\qquad\left.\times\exp\left[i\left(\underline{\kappa}^{(m,n)}-\alpha_1^{(m,n)}\underline{u}_{\mathrm{z}}\right)\boldsymbol{\cdot}\underline{r}\right]\right\}\\
\underline{H}_{\mathrm{ref}}(\underline{r}) &= in_1\eta_0^{-1}\sum_{m\in\mathbb{Z}}\sum_{n\in\mathbb{Z}}\left\{\left(r_{\mathrm{L}}^{(m,n)}\frac{i\underline{s}^{(m,n)}-\underline{p}_{\mathrm{ref}}^{(m,n)}}{\sqrt{2}}+r_{\mathrm{R}}^{(m,n)}\frac{i\underline{s}^{(m,n)}+\underline{p}_{\mathrm{ref}}^{(m,n)}}{\sqrt{2}}\right)\right.\\
&\qquad\left.\times\exp\left[i\left(\underline{\kappa}^{(m,n)}-\alpha_1^{(m,n)}\underline{u}_{\mathrm{z}}\right)\boldsymbol{\cdot}\underline{r}\right]\right\}\\
& \qquad\qquad\qquad\qquad z<0,
\end{aligned}\right\},\qquad(4.20)
$$

and

$$
\underline{E}_{tr}(\underline{r}) = \sum_{m\in\mathbb{Z}} \sum_{n\in\mathbb{Z}} \left\{ \left(t_L^{(m,n)} \frac{i\underline{s}^{(m,n)} - \underline{p}_{tr}^{(m,n)}}{\sqrt{2}} - t_R^{(m,n)} \frac{i\underline{s}^{(m,n)} + \underline{p}_{tr}^{(m,n)}}{\sqrt{2}} \right) \right.
$$

$$
\left. \times \exp\left[i \left(\underline{\kappa}^{(m,n)} + \alpha_2^{(m,n)} \underline{u}_z \right) \cdot (\underline{r} - d\,\underline{u}_z) \right] \right\}
$$

$$
\underline{H}_{tr}(\underline{r}) = -i n_2 \eta_0^{-1} \sum_{m\in\mathbb{Z}} \sum_{n\in\mathbb{Z}} \left\{ \left(t_L^{(m,n)} \frac{i\underline{s}^{(m,n)} - \underline{p}_{tr}^{(m,n)}}{\sqrt{2}} + t_R^{(m,n)} \frac{i\underline{s}^{(m,n)} + \underline{p}_{tr}^{(m,n)}}{\sqrt{2}} \right) \right.
$$

$$
\left. \times \exp\left[i \left(\underline{\kappa}^{(m,n)} + \alpha_2^{(m,n)} \underline{u}_z \right) \cdot (\underline{r} - d\,\underline{u}_z) \right] \right\}
$$

$$
z > d, \qquad (4.21)
$$

respectively. The amplitudes of the circularly polarized components in (4.19)–(4.21) are related to the amplitudes of the linearly polarized components in (4.7), (4.11), and (4.13) as follows:

$$
\left.\begin{aligned}
a_L^{(m,n)} &= -\left(i a_s^{(m,n)} + a_p^{(m,n)} \right) / \sqrt{2}, & a_R^{(m,n)} &= \left(i a_s^{(m,n)} - a_p^{(m,n)} \right) / \sqrt{2} \\
r_L^{(m,n)} &= \left(i r_s^{(m,n)} + r_p^{(m,n)} \right) / \sqrt{2}, & r_R^{(m,n)} &= -\left(i r_s^{(m,n)} - r_p^{(m,n)} \right) / \sqrt{2} \\
t_L^{(m,n)} &= -\left(i t_s^{(m,n)} + t_p^{(m,n)} \right) / \sqrt{2}, & t_R^{(m,n)} &= \left(i t_s^{(m,n)} - t_p^{(m,n)} \right) / \sqrt{2}
\end{aligned}\right\} . \qquad (4.22)
$$

In an analogous manner to (4.15), the circular reflection coefficients ($r_{LR}^{(m,n)}$, etc.) and the circular transmission coefficients ($t_{LR}^{(m,n)}$, etc.) of order (m, n) relate the reflection and transmission amplitudes of order (m, n) to the amplitudes of the incident plane wave per

$$
\left.\begin{aligned}
\begin{bmatrix} r_L^{(m,n)} \\ r_R^{(m,n)} \end{bmatrix} &= \begin{bmatrix} r_{LL}^{(m,n)} & r_{LR}^{(m,n)} \\ r_{RL}^{(m,n)} & r_{RR}^{(m,n)} \end{bmatrix} \begin{bmatrix} a_L^{(0,0)} \\ a_R^{(0,0)} \end{bmatrix} \\[2mm]
\begin{bmatrix} t_L^{(m,n)} \\ t_R^{(m,n)} \end{bmatrix} &= \begin{bmatrix} t_{LL}^{(m,n)} & t_{LR}^{(m,n)} \\ t_{RL}^{(m,n)} & t_{RR}^{(m,n)} \end{bmatrix} \begin{bmatrix} a_L^{(0,0)} \\ a_R^{(0,0)} \end{bmatrix}
\end{aligned}\right\} . \qquad (4.23)
$$

As previously discussed, cross-polarized coefficients have non-identical subscripts whereas co-polarized coefficients do not. Also, the superscript $^{(0,0)}$ identifies specular coefficients but not nonspecular coefficients. The circular reflection and transmission coefficients for each index pair (m, n) are related to the corresponding linear reflection and transmission coefficients via analogs of (3.39).

In the same vein as (4.16) and (4.17), circular reflectances of order (m,n) are defined as

$$R_{\mathrm{LR}}^{(m,n)} = \frac{\mathrm{Re}\left\{\alpha_1^{(m,n)}\right\}}{\alpha_1^{(0,0)}} |r_{\mathrm{LR}}^{(m,n)}|^2 ,$$

(4.24)

etc., and circular transmittances as

$$T_{\mathrm{LR}}^{(m,n)} = \frac{\mathrm{Re}\left\{\alpha_2^{(m,n)}\right\}}{\alpha_1^{(0,0)}} |t_{\mathrm{LR}}^{(m,n)}|^2 ,$$

(4.25)

etc. Thus, absorptances for left and right circular polarization states are defined as

$$\left.\begin{aligned}
A_{\mathrm{L}} &= 1 - \sum_{m\in\mathbb{Z}} \sum_{n\in\mathbb{Z}} \left(R_{\mathrm{LL}}^{(m,n)} + R_{\mathrm{RL}}^{(m,n)} + T_{\mathrm{LL}}^{(m,n)} + T_{\mathrm{RL}}^{(m,n)} \right) \\
A_{\mathrm{R}} &= 1 - \sum_{m\in\mathbb{Z}} \sum_{n\in\mathbb{Z}} \left(R_{\mathrm{RR}}^{(m,n)} + R_{\mathrm{LR}}^{(m,n)} + T_{\mathrm{RR}}^{(m,n)} + T_{\mathrm{LR}}^{(m,n)} \right)
\end{aligned}\right\} ,$$

(4.26)

respectively. If the materials of all three layers are passive then $0 \leq A_{\mathrm{L}} \leq 1$ and $0 \leq A_{\mathrm{R}} \leq 1$, with $A_{\mathrm{L}} = A_{\mathrm{R}} = 0$ arising only when all three materials are neither dissipative nor active [1].

4.6   RIGOROUS COUPLED-WAVE APPROACH

In order to determine amplitudes of the Floquet harmonics in the reflected and transmitted field phasors, the rigorous coupled-wave approach (RCWA) [14–18] is adopted. In this approach, all field phasors and constitutive dyadics are expressed as Fourier series with respect to both x and y.

The constitutive relations of the materials of layers $\mathcal{A}$, $\mathcal{B}$, and $\mathcal{C}$ can be jointly represented as

$$\left.\begin{aligned}
\underline{D}(\mathbf{r}) &= \underline{\underline{\varepsilon}}(x,y,z) \cdot \underline{E}(\mathbf{r}) + \underline{\underline{\xi}}(x,y,z) \cdot \underline{H}(\mathbf{r}) \\
\underline{B}(\mathbf{r}) &= \underline{\underline{\zeta}}(x,y,z) \cdot \underline{E}(\mathbf{r}) + \underline{\underline{\mu}}(x,y,z) \cdot \underline{H}(\mathbf{r})
\end{aligned}\right\} , \quad z \in (0,d) ,$$

(4.27)

with the nonhomogeneous 3×3 constitutive dyadics

$$\underline{\underline{\sigma}}(x,y,z) = \begin{cases}
\underline{\underline{\sigma}}^{\mathcal{A}} , & z \in (0, d_{\mathcal{A}} + g_{\min}] \\
\underline{\underline{\sigma}}^{\mathcal{A}} + \left(\underline{\underline{\sigma}}^{\mathcal{B}} - \underline{\underline{\sigma}}^{\mathcal{A}}\right) \mathcal{U}\left[z - d_{\mathcal{A}} - g(x,y)\right] , & z \in (d_{\mathcal{A}} + g_{\min}, d_{\mathcal{A}} + g_{\max}) \\
\underline{\underline{\sigma}}^{\mathcal{B}} , & z \in [d_{\mathcal{A}} + g_{\max}, d_{\mathcal{A}} + d_{\mathcal{B}}) \\
\underline{\underline{\sigma}}^{\mathcal{C}} , & z \in [d_{\mathcal{A}} + d_{\mathcal{B}}, d)
\end{cases} ,$$

(4.28)

wherein $\sigma \in \{\varepsilon, \xi, \zeta, \mu\}$ and the unit step function $\mathcal{U}(s)$ is defined as

$$\mathcal{U}(s) = \begin{cases} 1, & s \geq 0 \\ 0, & s < 0 \end{cases}. \tag{4.29}$$

Each of the four nonhomogeneous 3×3 constitutive dyadics is represented as a Fourier series per

$$\underline{\underline{\sigma}}(x, y, z) = \sum_{m \in \mathbb{Z}} \sum_{n \in \mathbb{Z}} \underline{\underline{\sigma}}^{(m,n)}(z) \exp\left[i 2\pi \left(m \frac{x}{L_x} + n \frac{y}{L_y}\right)\right], \quad z \in (0, d), \tag{4.30}$$

with dyadic Fourier coefficients

$$\underline{\underline{\sigma}}^{(m,n)}(z) = \begin{cases} \underline{\underline{\sigma}}^A \delta_{m0}\delta_{n0}, & z \in (0, d_A + g_{\min}] \\ \underline{\underline{\sigma}}^A \delta_{m0}\delta_{n0} + \left(\underline{\underline{\sigma}}^B - \underline{\underline{\sigma}}^A\right) \Upsilon^{(m,n)}(z), & z \in (d_A + g_{\min}, d_A + g_{\max}) \\ \underline{\underline{\sigma}}^B \delta_{m0}\delta_{n0}, & z \in [d_A + g_{\max}, d_A + d_B) \\ \underline{\underline{\sigma}}^C \delta_{m0}\delta_{n0}, & z \in [d_A + d_B, d) \end{cases}, \tag{4.31}$$

wherein the z-dependent scalar function

$$\Upsilon^{(m,n)}(z) = \frac{1}{L_x} \frac{1}{L_y} \int_0^{L_y} \int_0^{L_x} \mathcal{U}\left[z - d_A - g(x, y)\right] \exp\left[-i 2\pi \left(m \frac{x}{L_x} + n \frac{y}{L_y}\right)\right] dx \, dy. \tag{4.32}$$

In a similar vein, Fourier series representations are also introduced for the field phasors in the region $0 < z < d$ as follows:

$$\left. \begin{aligned} \underline{E}(\underline{r}) &= \sum_{m \in \mathbb{Z}} \sum_{n \in \mathbb{Z}} \left\{\underline{e}^{(m,n)}(z) \exp\left[i \underline{\kappa}^{(m,n)} \cdot \underline{r}\right]\right\} \\ \underline{H}(\underline{r}) &= \sum_{m \in \mathbb{Z}} \sum_{n \in \mathbb{Z}} \left\{\underline{h}^{(m,n)}(z) \exp\left[i \underline{\kappa}^{(m,n)} \cdot \underline{r}\right]\right\} \end{aligned} \right\}, \quad z \in (0, d). \tag{4.33}$$

The vector functions $\underline{e}^{(m,n)}(z)$ and $\underline{h}^{(m,n)}(z)$ herein are to be determined for all index pairs (m, n).

The summations on the right sides of (4.7), (4.11), (4.13), (4.19)–(4.21), (4.30), and (4.33) involve an infinite number of terms. The series involved are assumed to converge [19]. Thus, for practical purposes, the summations can be truncated after a suitably large number of terms. That is, the truncations $m \in [-M_t, M_t]$ and $n \in [-N_t, N_t]$ are imposed for these equations, with $M_t \geq 0$ and $N_t \geq 0$ being sufficiently large. Accordingly, a series containing $(2M_t + 1)(2N_t + 1)$ terms is implemented to represent every field phasor and every constitutive dyadic. For example, the reflected field phasors in (4.11) are recast as

$$\underline{E}_{\rm ref}(\underline{r}) = \sum_{m=-M_t}^{M_t} \sum_{n=-N_t}^{N_t} \left\{ \left(r_{\rm s}^{(m,n)}\, \underline{s}^{(m,n)} + r_{\rm p}^{(m,n)}\, \underline{p}_{\rm ref}^{(m,n)} \right) \right.$$
$$\left. \times \exp\left[i \left(\underline{\kappa}^{(m,n)} - \alpha_1^{(m,n)} \underline{u}_{\rm z} \right) \cdot \underline{r} \right] \right\}$$

$$\underline{H}_{\rm ref}(\underline{r}) = n_1 \eta_0^{-1} \sum_{m=-M_t}^{M_t} \sum_{n=-N_t}^{N_t} \left\{ \left(r_{\rm s}^{(m,n)}\, \underline{p}_{\rm ref}^{(m,n)} - r_{\rm p}^{(m,n)}\, \underline{s}^{(m,n)} \right) \right.$$
$$\left. \times \exp\left[i \left(\underline{\kappa}^{(m,n)} - \alpha_1^{(m,n)} \underline{u}_{\rm z} \right) \cdot \underline{r} \right] \right\}$$

$$\left. \right\}, \quad z < 0. \quad (4.34)$$

For compact presentation, the double summations over indexes m and n can be replaced by single summations over the superindex

$$\tau = m(2N_t + 1) + n, \quad m \in [-M_t, M_t], \quad n \in [-N_t, N_t], \quad (4.35)$$

with $\tau \in [-\tau_t, \tau_t]$ and $\tau_t = 2M_t N_t + M_t + N_t$. Thus, the reflected field phasors in (4.34) are compactly expressed as

$$\underline{E}_{\rm ref}(\underline{r}) = \sum_{\tau=-\tau_t}^{\tau_t} \left\{ \left(r_{\rm s}^{(\tau)}\, \underline{s}^{(\tau)} + r_{\rm p}^{(\tau)}\, \underline{p}_{\rm ref}^{(\tau)} \right) \exp\left[i \left(\underline{\kappa}^{(\tau)} - \alpha_1^{(\tau)} \underline{u}_{\rm z} \right) \cdot \underline{r} \right] \right\}$$

$$\underline{H}_{\rm ref}(\underline{r}) = n_1 \eta_0^{-1} \sum_{\tau=-\tau_t}^{\tau_t} \left\{ \left(r_{\rm s}^{(\tau)}\, \underline{p}_{\rm ref}^{(\tau)} - r_{\rm p}^{(\tau)}\, \underline{s}^{(\tau)} \right) \exp\left[i \left(\underline{\kappa}^{(\tau)} - \alpha_1^{(\tau)} \underline{u}_{\rm z} \right) \cdot \underline{r} \right] \right\}$$

$$\left. \right\}, \quad z < 0,$$

$$(4.36)$$

with the superscript $^{(\tau)}$ in (4.36) replacing the superscript pair $^{(m,n)}$ in (4.34).

The $(2\tau_t + 1)$-column vectors[1]

$$[\breve{\underline{e}}_\nu(z)] = \left[\breve{e}_\nu^{(-\tau_t)}(z),\ \breve{e}_\nu^{(-\tau_t+1)}(z),\ ...,\ \breve{e}_\nu^{(\tau_t-1)}(z),\ \breve{e}_\nu^{(\tau_t)}(z) \right]^T$$

$$\left[\breve{\underline{h}}_\nu(z) \right] = \left[\breve{h}_\nu^{(-\tau_t)}(z),\ \breve{h}_\nu^{(-\tau_t+1)}(z),\ ...,\ \breve{h}_\nu^{(\tau_t-1)}(z),\ \breve{h}_\nu^{(\tau_t)}(z) \right]^T$$

$$\left. \right\}, \quad \nu \in \{x, y, z\}, \quad (4.37)$$

are introduced for use in the region $0 \le z \le d$. The components of $[\breve{\underline{e}}_\nu(z)]$ and $\left[\breve{\underline{h}}_\nu(z) \right]$ are defined in terms of the Cartesian components of $\underline{e}^{(m,n)}(z)$ and $\underline{h}^{(m,n)}(z)$, respectively, per

$$\breve{e}_\nu^{(\tau)}(z) = e_\nu^{(m,n)}(z)$$
$$\breve{h}_\nu^{(\tau)}(z) = h_\nu^{(m,n)}(z)$$

$$\left. \right\}, \quad \nu \in \{x, y, z\}. \quad (4.38)$$

[1]The symbol $\breve{}$ identifies quantities associated with the RCWA.

As well as the column vectors defined in (4.37), for $\sigma \in \{\varepsilon, \xi, \zeta, \mu\}$, $\nu \in \{x, y, z\}$, and $\rho \in \{x, y, z\}$, $(2\tau_t + 1) \times (2\tau_t + 1)$ constitutive matrixes are defined as

$$
\left[\underline{\underline{\breve{\sigma}}}_{\nu\rho}(z)\right] = \begin{bmatrix} \breve{\sigma}_{\nu\rho}^{(-\tau_t,-\tau_t)}(z) & \breve{\sigma}_{\nu\rho}^{(-\tau_t,-\tau_t+1)}(z) & \cdots & \breve{\sigma}_{\nu\rho}^{(-\tau_t,\tau_t-1)}(z) & \breve{\sigma}_{\nu\rho}^{(-\tau_t,\tau_t)}(z) \\ \breve{\sigma}_{\nu\rho}^{(-\tau_t+1,-\tau_t)}(z) & \breve{\sigma}_{\nu\rho}^{(-\tau_t+1,-\tau_t+1)}(z) & \cdots & \breve{\sigma}_{\nu\rho}^{(-\tau_t+1,\tau_t-1)}(z) & \breve{\sigma}_{\nu\rho}^{(-\tau_t+1,\tau_t)}(z) \\ \cdots & \cdots & \cdots & \cdots & \cdots \\ \breve{\sigma}_{\nu\rho}^{(\tau_t-1,-\tau_t)}(z) & \breve{\sigma}_{\nu\rho}^{(\tau_t-1,-\tau_t+1)}(z) & \cdots & \breve{\sigma}_{\nu\rho}^{(\tau_t-1,\tau_t-1)}(z) & \breve{\sigma}_{\nu\rho}^{(\tau_t-1,\tau_t)}(z) \\ \breve{\sigma}_{\nu\rho}^{(\tau_t,-\tau_t)}(z) & \breve{\sigma}_{\nu\rho}^{(\tau_t,-\tau_t+1)}(z) & \cdots & \breve{\sigma}_{\nu\rho}^{(\tau_t,\tau_t-1)}(z) & \breve{\sigma}_{\nu\rho}^{(\tau_t,\tau_t)}(z) \end{bmatrix},
$$

$$(4.39)$$

with

$$
\breve{\sigma}_{\nu\rho}^{(\tau,\tau')}(z) = \sigma_{\nu\rho}^{(m-m',n-n')}(z) \tag{4.40}
$$

and

$$
\tau' = m'(2N_t + 1) + n', \quad m' \in [-M_t, M_t], \quad n' \in [-N_t, N_t]. \tag{4.41}
$$

Also, the $(2\tau_t + 1) \times (2\tau_t + 1)$ diagonal matrixes

$$
\left[\underline{\underline{\breve{K}}}_{\nu}\right] = \mathrm{diag}\left[\breve{k}_{\nu}^{(-\tau_t)}, \breve{k}_{\nu}^{(-\tau_t+1)}, ..., \breve{k}_{\nu}^{(\tau_t-1)}, \breve{k}_{\nu}^{(\tau_t)}\right], \qquad \nu \in \{x, y\}, \tag{4.42}
$$

comprising Fourier wavenumbers are defined using the scalars

$$
\left.\begin{aligned} \breve{k}_x^{(\tau)} &= k_x^{(m)} \\ \breve{k}_y^{(\tau)} &= k_y^{(n)} \end{aligned}\right\}. \tag{4.43}
$$

Equations (4.27)–(4.33) are now substituted in the Maxwell curl postulates $(2.16)_{1,2}$, under the presumption that $\underline{J}_{\text{ext}}(\underline{r}) = \underline{0}$ everywhere. The four matrix ordinary differential equations

$$
\frac{d}{dz}\left[\breve{e}_x(z)\right] = i\left[\underline{\underline{\breve{K}}}_x\right] \cdot \left[\breve{e}_z(z)\right] + i\omega\left\{\left[\underline{\underline{\breve{\zeta}}}_{yx}(z)\right] \cdot \left[\breve{e}_x(z)\right] + \left[\underline{\underline{\breve{\zeta}}}_{yy}(z)\right] \cdot \left[\breve{e}_y(z)\right]\right.
$$

$$
+ \left[\underline{\underline{\breve{\zeta}}}_{yz}(z)\right] \cdot \left[\breve{e}_z(z)\right] + \left[\underline{\underline{\breve{\mu}}}_{yx}(z)\right] \cdot \left[\breve{h}_x(z)\right] + \left[\underline{\underline{\breve{\mu}}}_{yy}(z)\right] \cdot \left[\breve{h}_y(z)\right]
$$

$$
\left. + \left[\underline{\underline{\breve{\mu}}}_{yz}(z)\right] \cdot \left[\breve{h}_z(z)\right]\right\}
$$

$$
\frac{d}{dz}\left[\breve{e}_y(z)\right] = i\left[\underline{\underline{\breve{K}}}_y\right] \cdot \left[\breve{e}_z(z)\right] - i\omega\left\{\left[\underline{\underline{\breve{\zeta}}}_{xx}(z)\right] \cdot \left[\breve{e}_x(z)\right] + \left[\underline{\underline{\breve{\zeta}}}_{xy}(z)\right] \cdot \left[\breve{e}_y(z)\right]\right.
$$

$$
+ \left[\underline{\underline{\breve{\zeta}}}_{xz}(z)\right] \cdot \left[\breve{e}_z(z)\right] + \left[\underline{\underline{\breve{\mu}}}_{xx}(z)\right] \cdot \left[\breve{h}_x(z)\right] + \left[\underline{\underline{\breve{\mu}}}_{xy}(z)\right] \cdot \left[\breve{h}_y(z)\right]
$$

$$
\left. + \left[\underline{\underline{\breve{\mu}}}_{xz}(z)\right] \cdot \left[\breve{h}_z(z)\right]\right\}
$$

$$
\frac{d}{dz}\left[\breve{h}_x(z)\right] = i\left[\underline{\underline{\breve{K}}}_x\right] \cdot \left[\breve{h}_z(z)\right] - i\omega\left\{\left[\underline{\underline{\breve{\varepsilon}}}_{yx}(z)\right] \cdot \left[\breve{e}_x(z)\right] + \left[\underline{\underline{\breve{\varepsilon}}}_{yy}(z)\right] \cdot \left[\breve{e}_y(z)\right]\right.
$$

$$
+ \left[\underline{\underline{\breve{\varepsilon}}}_{yz}(z)\right] \cdot \left[\breve{e}_z(z)\right] + \left[\underline{\underline{\breve{\xi}}}_{yx}(z)\right] \cdot \left[\breve{h}_x(z)\right] + \left[\underline{\underline{\breve{\xi}}}_{yy}(z)\right] \cdot \left[\breve{h}_y(z)\right]
$$

$$
\left. + \left[\underline{\underline{\breve{\xi}}}_{yz}(z)\right] \cdot \left[\breve{h}_z(z)\right]\right\}
$$

$$
\frac{d}{dz}\left[\breve{h}_y(z)\right] = i\left[\underline{\underline{\breve{K}}}_y\right] \cdot \left[\breve{h}_z(z)\right] + i\omega\left\{\left[\underline{\underline{\breve{\varepsilon}}}_{xx}(z)\right] \cdot \left[\breve{e}_x(z)\right] + \left[\underline{\underline{\breve{\varepsilon}}}_{xy}(z)\right] \cdot \left[\breve{e}_y(z)\right]\right.
$$

$$
+ \left[\underline{\underline{\breve{\varepsilon}}}_{xz}(z)\right] \cdot \left[\breve{e}_z(z)\right] + \left[\underline{\underline{\breve{\xi}}}_{xx}(z)\right] \cdot \left[\breve{h}_x(z)\right] + \left[\underline{\underline{\breve{\xi}}}_{xy}(z)\right] \cdot \left[\breve{h}_y(z)\right]
$$

$$
\left. + \left[\underline{\underline{\breve{\xi}}}_{xz}(z)\right] \cdot \left[\breve{h}_z(z)\right]\right\}
$$

$$
(4.44)
$$

and the two matrix algebraic equations

$$
\begin{aligned}
\left[\underline{\underline{\check{K}}}_x\right] \cdot [\underline{\check{e}}_y(z)] - \left[\underline{\underline{\check{K}}}_y\right] \cdot [\underline{\check{e}}_x(z)] = \omega & \left\{ \left[\underline{\underline{\check{\zeta}}}_{zx}(z)\right] \cdot [\underline{\check{e}}_x(z)] + \left[\underline{\underline{\check{\zeta}}}_{zy}(z)\right] \cdot [\underline{\check{e}}_y(z)] \right. \\
& + \left[\underline{\underline{\check{\zeta}}}_{zz}(z)\right] \cdot [\underline{\check{e}}_z(z)] + \left[\underline{\underline{\check{\mu}}}_{zx}(z)\right] \cdot \left[\underline{\check{h}}_x(z)\right] \\
& \left. + \left[\underline{\underline{\check{\mu}}}_{zy}(z)\right] \cdot \left[\underline{\check{h}}_y(z)\right] + \left[\underline{\underline{\check{\mu}}}_{zz}(z)\right] \cdot \left[\underline{\check{h}}_z(z)\right] \right\} \\[2ex]
\left[\underline{\underline{\check{K}}}_x\right] \cdot \left[\underline{\check{h}}_y(z)\right] - \left[\underline{\underline{\check{K}}}_y\right] \cdot \left[\underline{\check{h}}_x(z)\right] = -\omega & \left\{ \left[\underline{\underline{\check{\varepsilon}}}_{zx}(z)\right] \cdot [\underline{\check{e}}_x(z)] + \left[\underline{\underline{\check{\varepsilon}}}_{zy}(z)\right] \cdot [\underline{\check{e}}_y(z)] \right. \\
& + \left[\underline{\underline{\check{\varepsilon}}}_{zz}(z)\right] \cdot [\underline{\check{e}}_z(z)] + \left[\underline{\underline{\check{\xi}}}_{zx}(z)\right] \cdot \left[\underline{\check{h}}_x(z)\right] \\
& \left. + \left[\underline{\underline{\check{\xi}}}_{zy}(z)\right] \cdot \left[\underline{\check{h}}_y(z)\right] + \left[\underline{\underline{\check{\xi}}}_{zz}(z)\right] \cdot \left[\underline{\check{h}}_z(z)\right] \right\}
\end{aligned}
\tag{4.45}
$$

arise by taking advantage of the orthogonality properties of the functions

$$
\exp\left[i2\pi\left(m\frac{x}{L_x} + n\frac{y}{L_y}\right)\right], \quad m \in [-M_t, M_t], \quad n \in [-N_t, N_t], \tag{4.46}
$$

on the rectangular region $\{-L_x/2 \le x \le L_x/2, -L_y/2 \le y \le L_y/2\}$.

It is assumed that (4.45) can be solved for the column vectors $[\underline{\check{e}}_z(z)]$ and $\left[\underline{\check{h}}_z(z)\right]$, thereby enabling (4.44) to be compactly represented by a single matrix ordinary differential equation

$$
\frac{d}{dz}\left[\underline{\check{f}}(z)\right] = i\left[\underline{\underline{\check{P}}}(z)\right] \cdot \left[\underline{\check{f}}(z)\right], \quad z \in (0, d). \tag{4.47}
$$

Herein, the $4(2\tau_t + 1)$-column vector

$$
\left[\underline{\check{f}}(z)\right] = \begin{bmatrix} [\underline{\check{e}}_x(z)] \\ [\underline{\check{e}}_y(z)] \\ \left[\underline{\check{h}}_x(z)\right] \\ \left[\underline{\check{h}}_y(z)\right] \end{bmatrix}, \tag{4.48}
$$

while the $4(2\tau_t + 1) \times 4(2\tau_t + 1)$ matrix $\left[\underline{\underline{\check{P}}}(z)\right]$ is too unwieldy for display here.

Expressions for the boundary values $\left[\underline{\check{f}}(0-)\right]$ and $\left[\underline{\check{f}}(d+)\right]$ are provided by expansions of the incident and reflected field phasors, and expansions of the transmitted field phasors, respectively. Thus,

$$
\left[\breve{\underline{f}}(0-)\right] =
\begin{bmatrix}
\left[\breve{\underline{s}}_x\right] & \left[\breve{\underline{p}}^{\text{inc}}_x\right] & \left[\breve{\underline{s}}_x\right] & \left[\breve{\underline{p}}^{\text{ref}}_x\right] \\[6pt]
\left[\breve{\underline{s}}_y\right] & \left[\breve{\underline{p}}^{\text{inc}}_y\right] & \left[\breve{\underline{s}}_y\right] & \left[\breve{\underline{p}}^{\text{ref}}_y\right] \\[6pt]
n_1\eta_0^{-1}\left[\breve{\underline{p}}^{\text{inc}}_x\right] & -n_1\eta_0^{-1}\left[\breve{\underline{s}}_x\right] & n_1\eta_0^{-1}\left[\breve{\underline{p}}^{\text{ref}}_x\right] & -n_1\eta_0^{-1}\left[\breve{\underline{s}}_x\right] \\[6pt]
n_1\eta_0^{-1}\left[\breve{\underline{p}}^{\text{inc}}_y\right] & -n_1\eta_0^{-1}\left[\breve{\underline{s}}_y\right] & n_1\eta_0^{-1}\left[\breve{\underline{p}}^{\text{ref}}_y\right] & -n_1\eta_0^{-1}\left[\breve{\underline{s}}_y\right]
\end{bmatrix}
\cdot
\begin{bmatrix}
\left[\breve{\underline{A}}\right] \\[6pt]
\left[\breve{\underline{R}}\right]
\end{bmatrix},
$$

$$(4.49)$$

wherein the six $(2\tau_t + 1) \times (2\tau_t + 1)$ diagonal matrixes

$$
\left.
\begin{aligned}
\left[\breve{\underline{s}}_v\right] &=
\begin{bmatrix}
\underline{u}_v \cdot \underline{s}^{(-\tau_t)} & 0 & \cdots & 0 & 0 \\
0 & \underline{u}_v \cdot \underline{s}^{(-\tau_t+1)} & \cdots & 0 & 0 \\
\cdots & \cdots & \cdots & \cdots & \cdots \\
0 & 0 & \cdots & \underline{u}_v \cdot \underline{s}^{(\tau_t-1)} & 0 \\
0 & 0 & \cdots & 0 & \underline{u}_v \cdot \underline{s}^{(\tau_t)}
\end{bmatrix} \\[12pt]
\left[\breve{\underline{p}}^{\text{inc}}_v\right] &=
\begin{bmatrix}
\underline{u}_v \cdot \underline{p}^{(-\tau_t)}_{\text{inc}} & 0 & \cdots & 0 & 0 \\
0 & \underline{u}_v \cdot \underline{p}^{(-\tau_t+1)}_{\text{inc}} & \cdots & 0 & 0 \\
\cdots & \cdots & \cdots & \cdots & \cdots \\
0 & 0 & \cdots & \underline{u}_v \cdot \underline{p}^{(\tau_t-1)}_{\text{inc}} & 0 \\
0 & 0 & \cdots & 0 & \underline{u}_v \cdot \underline{p}^{(\tau_t)}_{\text{inc}}
\end{bmatrix} \\[12pt]
\left[\breve{\underline{p}}^{\text{ref}}_v\right] &=
\begin{bmatrix}
\underline{u}_v \cdot \underline{p}^{(-\tau_t)}_{\text{ref}} & 0 & \cdots & 0 & 0 \\
0 & \underline{u}_v \cdot \underline{p}^{(-\tau_t+1)}_{\text{ref}} & \cdots & 0 & 0 \\
\cdots & \cdots & \cdots & \cdots & \cdots \\
0 & 0 & \cdots & \underline{u}_v \cdot \underline{p}^{(\tau_t-1)}_{\text{ref}} & 0 \\
0 & 0 & \cdots & 0 & \underline{u}_v \cdot \underline{p}^{(\tau_t)}_{\text{ref}}
\end{bmatrix}
\end{aligned}
\right\}, \quad v \in \{x, y\}.
$$

$$(4.50)$$

Also, the $2(2\tau_t + 1)$-column vector

$$
\left[\breve{\underline{A}}\right] = \left[a_s^{(-\tau_t)}, a_s^{(-\tau_t+1)}, \dots, a_s^{(\tau_t-1)}, a_s^{(\tau_t)}, a_p^{(-\tau_t)}, a_p^{(-\tau_t+1)}, \dots, a_p^{(\tau_t-1)}, a_p^{(\tau_t)}\right]^T \quad (4.51)
$$

in (4.49) contains known incident phasor amplitudes, while the $2(2\tau_t + 1)$-column vector

$$
\left[\breve{\underline{R}}\right] = \left[r_s^{(-\tau_t)}, r_s^{(-\tau_t+1)}, \dots, r_s^{(\tau_t-1)}, r_s^{(\tau_t)}, r_p^{(-\tau_t)}, r_p^{(-\tau_t+1)}, \dots, r_p^{(\tau_t-1)}, r_p^{(\tau_t)}\right]^T \quad (4.52)
$$

contains unknown reflection phasor amplitudes.

Likewise,

$$
\left[\underline{\breve{f}}(d+)\right] =
\begin{bmatrix}
\left[\underline{\breve{s}}_x\right] & \left[\underline{\breve{p}}_x^{tr}\right] \\
\left[\underline{\breve{s}}_y\right] & \left[\underline{\breve{p}}_y^{tr}\right] \\
n_2\eta_0^{-1}\left[\underline{\breve{p}}_x^{tr}\right] & -n_2\eta_0^{-1}\left[\underline{\breve{s}}_x\right] \\
n_2\eta_0^{-1}\left[\underline{\breve{p}}_y^{tr}\right] & -n_2\eta_0^{-1}\left[\underline{\breve{s}}_y\right]
\end{bmatrix}
\cdot \left[\underline{\breve{T}}\right] ,
\tag{4.53}
$$

wherein the two $(2\tau_t + 1) \times (2\tau_t + 1)$ diagonal matrixes

$$
\left[\underline{\breve{p}}_v^{tr}\right] =
\begin{bmatrix}
\underline{u}_v \cdot \underline{p}_{tr}^{(-\tau_t)} & 0 & \dots & 0 & 0 \\
0 & \underline{u}_v \cdot \underline{p}_{tr}^{(-\tau_t+1)} & \dots & 0 & 0 \\
\dots & \dots & \dots & \dots & \dots \\
0 & 0 & \dots & \underline{u}_v \cdot \underline{p}_{tr}^{(\tau_t-1)} & 0 \\
0 & 0 & \dots & 0 & \underline{u}_v \cdot \underline{p}_{tr}^{(\tau_t)}
\end{bmatrix} ,
\quad v \in \{x, y\},
\tag{4.54}
$$

while the $2(2\tau_t + 1)$-column vector

$$
\left[\underline{\breve{T}}\right] = \left[t_s^{(-\tau_t)}, t_s^{(-\tau_t+1)}, \dots, t_s^{(\tau_t-1)}, t_s^{(\tau_t)}, t_p^{(-\tau_t)}, t_p^{(-\tau_t+1)}, \dots, t_p^{(\tau_t-1)}, t_p^{(\tau_t)}\right]^T
\tag{4.55}
$$

contains the unknown transmission phasor amplitudes.

The boundary values $\left[\underline{\breve{f}}(0-)\right]$ and $\left[\underline{\breve{f}}(d+)\right]$—provided in (4.49) and (4.53), respectively—are more conveniently expressed in terms of the $2(2\tau_t + 1) \times 2(2\tau_t + 1)$ matrixes

$$
\left[\underline{\breve{Y}}_e^{inc}\right] =
\begin{bmatrix}
\left[\underline{\breve{s}}_x\right] & \left[\underline{\breve{p}}_x^{inc}\right] \\
\left[\underline{\breve{s}}_y\right] & \left[\underline{\breve{p}}_y^{inc}\right]
\end{bmatrix} ,
\quad
\left[\underline{\breve{Y}}_h^{inc}\right] = n_1\eta_0^{-1}
\begin{bmatrix}
\left[\underline{\breve{p}}_x^{inc}\right] & -\left[\underline{\breve{s}}_x\right] \\
\left[\underline{\breve{p}}_y^{inc}\right] & -\left[\underline{\breve{s}}_y\right]
\end{bmatrix}
$$

$$
\left[\underline{\breve{Y}}_e^{ref}\right] =
\begin{bmatrix}
\left[\underline{\breve{s}}_x\right] & \left[\underline{\breve{p}}_x^{ref}\right] \\
\left[\underline{\breve{s}}_y\right] & \left[\underline{\breve{p}}_y^{ref}\right]
\end{bmatrix} ,
\quad
\left[\underline{\breve{Y}}_h^{ref}\right] = n_1\eta_0^{-1}
\begin{bmatrix}
\left[\underline{\breve{p}}_x^{ref}\right] & -\left[\underline{\breve{s}}_x\right] \\
\left[\underline{\breve{p}}_y^{ref}\right] & -\left[\underline{\breve{s}}_y\right]
\end{bmatrix}
\tag{4.56}
$$

$$
\left[\underline{\breve{Y}}_e^{tr}\right] =
\begin{bmatrix}
\left[\underline{\breve{s}}_x\right] & \left[\underline{\breve{p}}_x^{tr}\right] \\
\left[\underline{\breve{s}}_y\right] & \left[\underline{\breve{p}}_y^{tr}\right]
\end{bmatrix} ,
\quad
\left[\underline{\breve{Y}}_h^{tr}\right] = n_2\eta_0^{-1}
\begin{bmatrix}
\left[\underline{\breve{p}}_x^{tr}\right] & -\left[\underline{\breve{s}}_x\right] \\
\left[\underline{\breve{p}}_y^{tr}\right] & -\left[\underline{\breve{s}}_y\right]
\end{bmatrix}
$$

as

$$
\left[\underline{\breve{f}}(0-)\right] =
\begin{bmatrix}
\left[\underline{\breve{Y}}_e^{inc}\right] & \left[\underline{\breve{Y}}_e^{ref}\right] \\
\left[\underline{\breve{Y}}_h^{inc}\right] & \left[\underline{\breve{Y}}_h^{ref}\right]
\end{bmatrix}
\cdot
\begin{bmatrix}
\left[\underline{\breve{A}}\right] \\
\left[\underline{\breve{R}}\right]
\end{bmatrix}
\tag{4.57}
$$

and

$$\left[\underline{\breve{f}}(d+)\right] = \left[\begin{array}{c} \left[\underline{\underline{\breve{Y}}}_e^{tr}\right] \\ \left[\underline{\underline{\breve{Y}}}_h^{tr}\right] \end{array}\right] \cdot \left[\underline{\breve{T}}\right] . \tag{4.58}$$

The appropriate RCWA machinery is now in place to formulate the the solution of (4.47). Due to the z-dependency of $\left[\underline{\underline{\breve{P}}}(z)\right]$ therein, a numerical approach is needed. The piecewise-uniform approximation method provides a suitable basis for calculations as follows: The region occupied by the layers $\mathcal{A}$, $\mathcal{B}$, and $\mathcal{C}$, i.e., $0 \leq z \leq d$, is divided into N_s subregions. The ℓ-th subregion is bounded by the planes $z = z_{\ell-1}$ and $z = z_\ell$, $\ell \in [1, N_s]$, where $z_0 = 0$ and $z_{N_s} = d$. Specifically, the region $0 \leq z \leq d_\mathcal{A} + g_{min}$ is divided into N_I subregions; the region $d_\mathcal{A} + g_{min} \leq z \leq d_\mathcal{A} + g_{max}$ is divided into N_{II} subregions; the region $d_\mathcal{A} + g_{max} \leq z \leq d_\mathcal{A} + d_\mathcal{B}$ is divided into N_{III} subregions; and the region $d_\mathcal{A} + d_\mathcal{B} \leq z \leq d$ is divided into N_{IV} subregions, with $N_s = N_I + N_{II} + N_{III} + N_{IV} > 4$.

For $z \in (0, d_\mathcal{A} + g_{min}] \cup (d_\mathcal{A} + g_{max}, d]$, within each subregion the matrix $\left[\underline{\underline{\breve{P}}}(z)\right]$ is independent of z and is simply given by the uniform matrix

$$\left[\underline{\underline{\breve{P}}}\right]^{(\ell)} = \left[\underline{\underline{\breve{P}}}(z_\ell)\right], \qquad \ell \in \{1, \ldots, N_I\} \cup \{N_I + N_{II} + 1, \ldots, N_s\} . \tag{4.59}$$

For $z \in (d_\mathcal{A} + g_{min}, d_\mathcal{A} + g_{max}]$, within each subregion the matrix $\left[\underline{\underline{\breve{P}}}(z)\right]$ is approximated by the uniform matrix

$$\left[\underline{\underline{\breve{P}}}\right]^{(\ell)} = \left[\underline{\underline{\breve{P}}}\left(\frac{z_\ell + z_{\ell-1}}{2}\right)\right], \qquad \ell \in \{N_I + 1, \ldots, N_I + N_{II}\} , \tag{4.60}$$

provided that every subregion is sufficiently thin. Then,

$$\left[\underline{\breve{f}}(d-)\right] \simeq \left[\underline{\underline{\breve{M}}}\right]^{(N_s)} \cdot \left[\underline{\underline{\breve{M}}}\right]^{(N_s-1)} \cdot \cdots \cdot \left[\underline{\underline{\breve{M}}}\right]^{(2)} \cdot \left[\underline{\underline{\breve{M}}}\right]^{(1)} \cdot \left[\underline{\breve{f}}(0+)\right] , \tag{4.61}$$

where

$$\left[\underline{\underline{\breve{M}}}\right]^{(\ell)} = \exp\left\{i(z_\ell - z_{\ell-1})\left[\underline{\underline{\breve{P}}}\right]^{(\ell)}\right\} , \qquad \ell \in [1, N_s] \tag{4.62}$$

is the *transfer matrix* of the ℓ-th subregion. Generally, N_I, N_{III}, and N_{IV} are simply unity. However, on occasion, especially when electrically thick regions are involved, it is helpful to increase the values of N_I, N_{III}, and N_{IV} for reasons of computational stability.

The standard boundary conditions imposed at $z = 0$ and $z = d$ are

$$\left.\begin{array}{c} \left[\underline{\breve{f}}(0-)\right] = \left[\underline{\breve{f}}(0+)\right] \\ \left[\underline{\breve{f}}(d-)\right] = \left[\underline{\breve{f}}(d+)\right] \end{array}\right\} . \tag{4.63}$$

The combination of (4.57), (4.58), and (4.61) yields the matrix equation

$$
\left[\begin{array}{c} \left[\underline{\underline{\check{Y}}}_e^{\,\mathrm{tr}} \right] \\ \left[\underline{\underline{\check{Y}}}_h^{\,\mathrm{tr}} \right] \end{array} \right] \cdot \left[\underline{\check{T}} \right] = \left[\underline{\underline{\check{M}}} \right]^{(N_s)} \cdot \left[\underline{\underline{\check{M}}} \right]^{(N_s-1)} \cdot \cdots
$$

$$
\cdots \cdot \left[\underline{\underline{\check{M}}} \right]^{(2)} \cdot \left[\underline{\underline{\check{M}}} \right]^{(1)} \cdot \left[\begin{array}{cc} \left[\underline{\underline{\check{Y}}}_e^{\,\mathrm{inc}} \right] & \left[\underline{\underline{\check{Y}}}_e^{\,\mathrm{ref}} \right] \\ \left[\underline{\underline{\check{Y}}}_h^{\,\mathrm{inc}} \right] & \left[\underline{\underline{\check{Y}}}_h^{\,\mathrm{ref}} \right] \end{array} \right] \cdot \left[\begin{array}{c} \left[\underline{\check{A}} \right] \\ \left[\underline{\check{R}} \right] \end{array} \right] , \qquad (4.64)
$$

from which $\left[\underline{\check{R}} \right]$ and $\left[\underline{\check{T}} \right]$ can be extracted via standard numerical techniques [20].

4.7   STABLE RCWA ALGORITHM

In principle, standard matrix-inversion techniques—for example, the Gauss elimination technique [20]—can be employed to solve (4.64). However, in practice, the implementation of such techniques can be bedeviled by numerical problems, especially if d is electrically large and/or the incidence with respect to the z axis is highly oblique. A stable RCWA algorithm—based on the diagonalization of the matrixes $\left[\underline{\underline{\check{P}}} \right]^{(\ell)}$ for each $\ell \in [1, N_s]$—has been developed to overcome these problems [15, 16, 21, 22]. A description of the stable RCWA algorithm follows.

It is assumed that $\left[\underline{\underline{\check{P}}} \right]^{(\ell)}$ is diagonalizable [23] for all $\ell \in [1, N_s]$. Accordingly, each matrix $\left[\underline{\underline{\check{P}}} \right]^{(\ell)}$ can expressed as

$$
\left[\underline{\underline{\check{P}}} \right]^{(\ell)} = \left[\underline{\underline{\check{V}}} \right]^{(\ell)} \cdot \left[\underline{\underline{\check{G}}} \right]^{(\ell)} \cdot \left(\left[\underline{\underline{\check{V}}} \right]^{(\ell)} \right)^{-1} , \qquad \ell \in [1, N_s] , \qquad (4.65)
$$

wherein the diagonal $4(2\tau_t + 1) \times 4(2\tau_t + 1)$ matrix $\left[\underline{\underline{\check{G}}} \right]^{(\ell)}$ contains the eigenvalues of $\left[\underline{\underline{\check{P}}} \right]^{(\ell)}$ in decreasing order of the magnitude of their imaginary parts, and the columns of the $4(2\tau_t + 1)$ $\times 4(2\tau_t + 1)$ matrix $\left[\underline{\underline{\check{V}}} \right]^{(\ell)}$ are the eigenvectors of $\left[\underline{\underline{\check{P}}} \right]^{(\ell)}$, with the order of each eigenvector in $\left[\underline{\underline{\check{V}}} \right]^{(\ell)}$ matching the order of the corresponding eigenvalue in $\left[\underline{\underline{\check{G}}} \right]^{(\ell)}$. Then, by solving (4.47), $\left[\underline{\check{f}}(z_{\ell-1}) \right]$ is related to $\left[\underline{\check{f}}(z_\ell) \right]$ by

$$
\left[\underline{\check{f}}(z_{\ell-1}) \right] = \left[\underline{\underline{\check{V}}} \right]^{(\ell)} \cdot \exp \left\{ -i\,(z_\ell - z_{\ell-1}) \left[\underline{\underline{\check{G}}} \right]^{(\ell)} \right\} \cdot \left(\left[\underline{\underline{\check{V}}} \right]^{(\ell)} \right)^{-1} \cdot \left[\underline{\check{f}}(z_\ell) \right] . \qquad (4.66)
$$

Next, it is assumed that a set of auxiliary $2(2\tau_t + 1)$-column vectors $\left[\check{T}\right]^{(\ell)}$ and auxiliary transmission matrixes $\left[\underline{\underline{\check{Z}}}\right]^{(\ell)}$ of size $4(2\tau_t + 1) \times 2(2\tau_t + 1)$ exist, such that [21]

$$\left[\underline{\check{f}}(z_\ell)\right] = \left[\underline{\underline{\check{Z}}}\right]^{(\ell)} \cdot \left[\check{T}\right]^{(\ell)}, \qquad \ell \in [0, N_s], \tag{4.67}$$

wherein

$$\left.\begin{array}{l} \left[\check{T}\right]^{(N_s)} = \left[\check{T}\right] \\[2mm] \left[\underline{\underline{\check{Z}}}\right]^{(N_s)} = \left[\begin{array}{c} \left[\underline{\underline{\check{Y}}}_e^{\mathrm{tr}}\right] \\[2mm] \left[\underline{\underline{\check{Y}}}_h^{\mathrm{tr}}\right] \end{array}\right] \end{array}\right\}. \tag{4.68}$$

The existence of $\left[\check{T}\right]^{(\ell)}$ and $\left[\underline{\underline{\check{Z}}}\right]^{(\ell)}$ per (4.67) enables (4.66) to be recast as

$$\left[\underline{\underline{\check{Z}}}\right]^{(\ell-1)} \cdot \left[\check{T}\right]^{(\ell-1)} =$$

$$\left[\underline{\underline{\check{V}}}\right]^{(\ell)} \cdot \left[\begin{array}{cc} \exp\left\{-i(z_\ell - z_{\ell-1})\left[\underline{\underline{\check{G}}}_{\mathrm{upper}}\right]^{(\ell)}\right\} & \left[\underline{\underline{\check{0}}}\right] \\[3mm] \left[\underline{\underline{\check{0}}}\right] & \exp\left\{-i(z_\ell - z_{\ell-1})\left[\underline{\underline{\check{G}}}_{\mathrm{lower}}\right]^{(\ell)}\right\} \end{array}\right]$$

$$\cdot \left(\left[\underline{\underline{\check{V}}}\right]^{(\ell)}\right)^{-1} \cdot \left[\underline{\underline{\check{Z}}}\right]^{(\ell)} \cdot \left[\check{T}\right]^{(\ell)}, \qquad \ell \in [1, N_s], \tag{4.69}$$

with the $2(2\tau_t + 1) \times 2(2\tau_t + 1)$ matrixes $\left[\underline{\underline{\check{G}}}_{\mathrm{upper}}\right]^{(\ell)}$ and $\left[\underline{\underline{\check{G}}}_{\mathrm{lower}}\right]^{(\ell)}$ being the upper and lower diagonal submatrixes of the $4(2\tau_t + 1) \times 4(2\tau_t + 1)$ matrix $\left[\underline{\underline{\check{G}}}\right]^{(\ell)}$, respectively, and $\left[\underline{\underline{\check{0}}}\right]$ being the $2(2\tau_t + 1) \times 2(2\tau_t + 1)$ null matrix.

A further two $2(2\tau_t + 1) \times 2(2\tau_t + 1)$ matrixes, namely $\left[\underline{\underline{\check{X}}}_{\mathrm{upper}}\right]^{(\ell)}$ and $\left[\underline{\underline{\check{X}}}_{\mathrm{lower}}\right]^{(\ell)}$, are introduced. These are defined by the relation

$$\left[\begin{array}{c} \left[\underline{\underline{\check{X}}}_{\mathrm{upper}}\right]^{(\ell)} \\[2mm] \left[\underline{\underline{\check{X}}}_{\mathrm{lower}}\right]^{(\ell)} \end{array}\right] = \left(\left[\underline{\underline{\check{V}}}\right]^{(\ell)}\right)^{-1} \cdot \left[\underline{\underline{\check{Z}}}\right]^{(\ell)}. \tag{4.70}$$

Also, it is postulated that the recurrence relation

$$\left[\check{T}\right]^{(\ell-1)} = \exp\left\{-i(z_\ell - z_{\ell-1})\left[\underline{\underline{\check{G}}}_{\mathrm{upper}}\right]^{(\ell)}\right\} \cdot \left[\underline{\underline{\check{X}}}_{\mathrm{upper}}\right]^{(\ell)} \cdot \left[\check{T}\right]^{(\ell)} \tag{4.71}$$

holds. Inversion of the recurrence relation (4.71) followed by substitution in (4.69) yields the relation

$$\left[\underline{\underline{\check{Z}}}\right]^{(\ell-1)} = \left[\underline{\underline{\check{V}}}\right]^{(\ell)} \cdot \begin{bmatrix} \left[\underline{\underline{\check{I}}}\right] \\ \left[\underline{\underline{\check{U}}}\right]^{(\ell)} \end{bmatrix}, \qquad \ell \in [1, N_s] , \tag{4.72}$$

wherein the $2(2\tau_t + 1) \times 2(2\tau_t + 1)$ matrix

$$\begin{aligned} \left[\underline{\underline{\check{U}}}\right]^{(\ell)} &= \exp\left\{-i(z_\ell - z_{\ell-1})\left[\underline{\underline{\check{G}}}_{\text{lower}}\right]^{(\ell)}\right\} \cdot \left[\underline{\underline{\check{X}}}_{\text{lower}}\right]^{(\ell)} \cdot \left(\left[\underline{\underline{\check{X}}}_{\text{upper}}\right]^{(\ell)}\right)^{-1} \\ &\cdot \exp\left\{i(z_\ell - z_{\ell-1})\left[\underline{\underline{\check{G}}}_{\text{upper}}\right]^{(\ell)}\right\}, \qquad \ell \in [1, N_s] , \end{aligned} \tag{4.73}$$

and $\left[\underline{\underline{\check{I}}}\right]$ is the $2(2\tau_t + 1) \times 2(2\tau_t + 1)$ identity matrix. Thus, by repeatedly applying (4.72) and (4.70), $\left[\underline{\underline{\check{Z}}}\right]^{(\ell)}$ can be obtained in terms of $\left[\underline{\underline{\check{Z}}}\right]^{(N_s)}$ $\forall \ell \in [0, N_s - 1]$.

It is helpful to partition $\left[\underline{\underline{\check{Z}}}\right]^{(0)}$ as

$$\left[\underline{\underline{\check{Z}}}\right]^{(0)} = \begin{bmatrix} \left[\underline{\underline{\check{Z}}}_{\text{upper}}\right]^{(0)} \\ \left[\underline{\underline{\check{Z}}}_{\text{lower}}\right]^{(0)} \end{bmatrix}. \tag{4.74}$$

The right sides of (4.57) and (4.67) evaluated for $\ell = 0$ can be equated, in consequence of the boundary condition $\left[\underline{\check{f}}(0-)\right] = \left[\underline{\check{f}}(0+)\right]$, to obtain

$$\begin{bmatrix} \left[\underline{\underline{\check{Z}}}_{\text{upper}}\right]^{(0)} \\ \left[\underline{\underline{\check{Z}}}_{\text{lower}}\right]^{(0)} \end{bmatrix} \cdot \left[\underline{\underline{\check{T}}}\right]^{(0)} = \begin{bmatrix} \left[\underline{\underline{\check{Y}}}_e^{\text{inc}}\right] & \left[\underline{\underline{\check{Y}}}_e^{\text{ref}}\right] \\ \left[\underline{\underline{\check{Y}}}_h^{\text{inc}}\right] & \left[\underline{\underline{\check{Y}}}_h^{\text{ref}}\right] \end{bmatrix} \cdot \begin{bmatrix} \left[\underline{\check{A}}\right] \\ \left[\underline{\check{R}}\right] \end{bmatrix}. \tag{4.75}$$

Hence, $\left[\underline{\check{R}}\right]$ and $\left[\underline{\underline{\check{T}}}\right]^{(0)}$ are given as

$$\begin{bmatrix} \left[\underline{\underline{\check{T}}}\right]^{(0)} \\ \left[\underline{\check{R}}\right] \end{bmatrix} = \begin{bmatrix} \left[\underline{\underline{\check{Z}}}_{\text{upper}}\right]^{(0)} & -\left[\underline{\underline{\check{Y}}}_e^{\text{ref}}\right] \\ \left[\underline{\underline{\check{Z}}}_{\text{lower}}\right]^{(0)} & -\left[\underline{\underline{\check{Y}}}_h^{\text{ref}}\right] \end{bmatrix}^{-1} \cdot \begin{bmatrix} \left[\underline{\underline{\check{Y}}}_e^{\text{inc}}\right] \\ \left[\underline{\underline{\check{Y}}}_h^{\text{inc}}\right] \end{bmatrix} \cdot \left[\underline{\check{A}}\right] . \tag{4.76}$$

Furthermore, once $\left[\underline{\underline{\check{T}}}\right]^{(0)}$ has been found, $\left[\underline{\underline{\check{T}}}\right]^{(\ell)}$ can be determined for all $\ell \in [1, N_s]$ via inversion of the recurrence relation (4.71). Thereby, both $\left[\underline{\check{T}}\right] = \left[\underline{\underline{\check{T}}}\right]^{(N_s)}$ and $\left[\underline{\check{R}}\right]$ are provided by the stable RCWA algorithm.

With $\left[\breve{\underline{T}}\right]$ and $\left[\breve{\underline{R}}\right]$ known, and under the assumption that $\left[\breve{\underline{A}}\right]$ is given, $\left[\breve{\underline{f}}(z_0)\right] = \left[\breve{\underline{f}}(0-)\right]$ can be found using (4.57). Thus, $\left[\breve{\underline{f}}(z_\ell)\right]$ for all $\ell \in [1, N_s]$ can be calculated by repeatedly applying (4.66). Then, with $\left[\breve{\underline{f}}(z_\ell)\right]$ known, $\left[\breve{\underline{h}}_z(z_\ell)\right]$ and $[\breve{\underline{e}}_z(z_\ell)]$ can be extracted from (4.45). Finally, the field phasors $\underline{E}(\underline{r})$ and $\underline{H}(\underline{r})$ throughout the region $0 \leq z \leq d$ are delivered by the application of (4.33).

How large should M_t and N_t be? This issue requires numerical experimentation: while increasing M_t and/or N_t in small increments, the convergence of the reflectances $R_{\text{sp}}^{(m,n)}$, etc., and the transmittances $T_{\text{sp}}^{(m,n)}$, etc. should be checked, until satisfactory tolerance limits (say, 0.1%) are achieved. The principle of conservation of energy must be satisfied simultaneously for all values of M_t and N_t, which can provide a useful check on the stability of computations.

4.8   REFERENCES

[1] Mackay, T. G. and Lakhtakia, A. 2019. *Electromagnetic Anisotropy and Bianisotropy: A Field Guide*, 2nd ed. (Singapore, World Scientific). DOI: 10.1142/11351. 51, 56, 59

[2] Boardman, A. D., Ed., 1982. *Electromagnetic Surface Modes* (Chicester, UK, Wiley). 51

[3] Polo, J. A., Jr., Mackay, T. G., and Lakhtakia, A. 2013. *Electromagnetic Surface Waves: A Modern Perspective* (Waltham, MA, Elsevier). 51, 52, 53, 54

[4] Lakhtakia, Λ. and Messier, R. 2005. *Sculptured Thin Films: Nanoengineered Morphology and Optics* (Bellingham, WA, SPIE Press). DOI: 10.1117/3.585322. 51

[5] Lakhtakia, A. and Polo, J. A., Jr. 2007. Dyakonov–Tamm wave at the planar interface of a chiral sculptured thin film and an isotropic dielectric material, *J. Eur. Opt. Soc.—Rapid Pub.*, 2:07021. DOI: 10.2971/jeos.2007.07021. 51

[6] Pulsifer, D. P., Faryad, M., and Lakhtakia, A. 2013. Observation of the Dyakonov–Tamm wave, *Phys. Rev. Lett.*, 111:243902. DOI: 10.1103/physrevlett.111.243902. 51

[7] Maystre, D., Ed., 1993. *Selected Papers on Diffraction Gratings* (Bellingham, WA, SPIE Press). 52

[8] Solano, M., Faryad, M., Hall, A. S., Mallouk, T. E., Monk, P. B., and Lakhtakia, A. 2013. Optimization of the absorption efficiency of an amorphous-silicon thin-film tandem solar cell backed by a metallic surface-relief grating, *Appl. Opt.*, 52:966–979. DOI: 10.1364/AO.52.000966.
Solano, M., Faryad, M., Hall, A. S., Mallouk, T. E., Monk, P. B., and Lakhtakia, A. 2015. Optimization of the absorption efficiency of an amorphous-silicon thin-film tandem solar cell backed by a metallic surface-relief grating, *Appl. Opt.*, 54:398–399. (erratum) DOI: 10.1364/AO.54.000398. 52

[9] Ahmad, F., Anderson, T. H., Civiletti, B. J., Monk, P. B., and Lakhtakia, A. 2018. On optical-absorption peaks in a nonhomogeneous thin-film solar cell with a two-dimensional periodically corrugated metallic backreflector, *J. Nanophoton.*, 12:016017. DOI: 10.1117/1.jnp.12.016017. 52

[10] Civiletti, B. J., Anderson, T. H., Ahmad, F., Monk, P. B. and Lakhtakia, A. 2018. Optimization approach for optical absorption in three-dimensional structures including solar cells, *Opt. Eng.*, 57:057101. DOI: 10.1117/1.oe.57.5.057101. 52

[11] Dolev, I., Volodarsky, M., Porat, G., and Arie, A. 2011. Multiple coupling of surface plasmons in quasiperiodic gratings, *Opt. Lett.*, 36:1584–1586. DOI: 10.1364/ol.36.001584. 52

[12] Faryad, M. and Lakhtakia, A. 2012. Excitation of multiple surface-plasmon-polariton waves using a compound surface-relief grating, *J. Nanophoton.*, 6:061701. DOI: 10.1117/1.jnp.6.061701. 52

[13] Lakhtakia, A., Varadan, V. K., and Varadan, V. V. 1985. Scattering by a partially illuminated, doubly periodic, doubly infinite surface, *J. Acoust. Soc. Am.*, 77:1999–2004. DOI: 10.1121/1.2022331. 53, 54

[14] Glytsis, E. N. and Gaylord, T. K. 1987. Rigorous three-dimensional coupled-wave diffraction analysis of single and cascaded anisotropic gratings, *J. Opt. Soc. Am. A*, 4:2061–2080. DOI: 10.1364/josaa.4.002061. 59

[15] Moharam, M. G. and Gaylord, T. K. 1981. Rigorous coupled-wave analysis of planar-grating diffraction, *J. Opt. Soc. Am.*, 71:811–818. DOI: 10.1364/josa.71.000811. 59, 68

[16] Chateau, N. and Hugonin, J.-P. 1994. Algorithm for the rigorous coupled-wave analysis of grating diffraction, *J. Opt. Soc. Am. A*, 11:1321–1331. DOI: 10.1364/josaa.11.001321. 59, 68

[17] Wang, F. and Lakhtakia, A. 2004. Lateral shifts of optical beams on reflection by slanted chiral sculptured thin films, *Opt. Commun.*, 235:107–132. DOI: 10.1016/j.optcom.2004.02.050. 59

[18] Onishi, M., Crabtree, K., and Chipman, R. A. 2011. Formulation of rigorous coupled-wave theory for gratings in bianisotropic media, *J. Opt. Soc. Am. A*, 28:1747–1758. DOI: 10.1364/josaa.28.001747. 59

[19] Lakhtakia, A., Varadan, V. K., and Varadan, V. V. 1991. On filling up the grooves of a perfectly-conducting grating with a dielectric material, *J. Modern Opt.*, 38:659–667. DOI: 10.1080/09500349114550661. 60

[20] Jaluria, Y. 1996. *Computer Methods for Engineering* (Washington, DC, Taylor & Francis). 68

[21] Wang, F., Horn, M. W., and Lakhtakia, A. 2004. Rigorous electromagnetic modeling of near-field phase-shifting contact lithography, *Microelectron. Eng.*, 71:34–53. DOI: 10.1016/j.mee.2003.09.003. 68, 69

[22] Li, L. 1993. Multilayer modal method for diffraction gratings of arbitrary profile, depth, and permittivity, *J. Opt. Soc. Am. A*, 10:2581–2591. DOI: 10.1364/JOSAA.10.002581. Li, L. 1994. Multilayer modal method for diffraction gratings of arbitrary profile, depth, and permittivity, *J. Opt. Soc. Am. A*, 11:1685 (addendum). DOI: 10.1364/JOSAA.11.001685. 68

[23] Hochstadt, H. 1975. *Differential Equations: A Modern Approach* (New York, Dover Press). 68

CHAPTER 5

Isotropic Dielectric Slab

In Chapters 3 and 4, the TMM is presented for the most general cases involving multilayered slabs of homogeneous bianisotropic materials with planar and periodically corrugated interfaces. The powerful formalisms presented therein can be extended to multilayered whose constituent layers are composed of materials that are nonhomogeneous in the thickness direction, by means of a piecewise-uniform approximation [1, 2].

Most commonly, the TMM is applied to a multilayered slab comprising layers made of isotropic dielectric materials, the constituent layers have either planar interfaces or singly periodic interfaces along the x axis, and the incident plane wave propagates wholly in the xz plane (i.e., $\psi = 0$). This chapter presents the TMM for this scenario.

5.1   MULTILAYERED SLAB WITH PLANAR INTERFACES

Consider the reflection–transmission problem schematically illustrated in Fig. 3.1. That is, a plane wave from a distant source is incident on a multilayered slab comprising three constituent layers, labeled $\mathcal{A}$, $\mathcal{B}$, and $\mathcal{C}$, each of infinite extent in the xy plane. The surfaces of the layers are all planar. Layer $\mathcal{A}$ has thickness d_A, layer $\mathcal{B}$ has thickness d_B, and layer $\mathcal{C}$ has thickness d_C. Each layer is made of a different homogeneous isotropic dielectric material. The constitutive relations of the materials that occupy the region $0 < z < d$ are

$$\left. \begin{array}{l} \underline{D}(\underline{r}) = \left\{ \begin{array}{ll} \varepsilon^A \underline{E}(\underline{r}), & 0 < z < d_A \\[2mm] \varepsilon^B \underline{E}(\underline{r}), & d_A < z < d_A + d_B \\[2mm] \varepsilon^C \underline{E}(\underline{r}), & d_A + d_B < z < d \end{array} \right. \\[8mm] \underline{B}(\underline{r}) = \mu_0 \underline{H}(\underline{r}), \qquad 0 < z < d \end{array} \right\}, \tag{5.1}$$

with ε^A, ε^B, and ε^C being the permittivities of the materials constituting layers $\mathcal{A}$, $\mathcal{B}$, and $\mathcal{C}$, respectively, and $d = d_A + d_B + d_C$. As in Section 3.1, the half-space $z < 0$ is occupied by a homogeneous isotropic dielectric material of refractive index $n_1 > 0$ and the half-space $z > d$ is occupied by a homogeneous isotropic dielectric material of refractive index $n_2 > 0$. The incident plane wave propagates wholly in the xz plane.

Expressions for the incident, reflected and transmitted field phasors, in terms of s- and p-polarized components, are precisely those in (3.4), (3.7), and (3.8), respectively, with $\psi = 0$. These phasors are required to satisfy the matrix ordinary differential equations (3.11), wherein the matrixes $\left[\underline{\underline{P}}\right]^{\gamma}$ are given as

$$\left[\underline{\underline{P}}\right]^{\gamma} = \begin{bmatrix} 0 & 0 & \omega\mu_0 - \dfrac{q^2}{\omega\varepsilon^{\gamma}} \\ 0 & 0 & -\omega\mu_0 & 0 \\ 0 & -\omega\varepsilon^{\gamma} + \dfrac{q^2}{\omega\mu_0} & 0 & 0 \\ \omega\varepsilon^{\gamma} & 0 & 0 \end{bmatrix}, \quad \gamma \in \{\mathcal{A}, \mathcal{B}, \mathcal{C}\}. \quad (5.2)$$

Thus, the corresponding transfer matrixes, as defined in (3.13), have the form

$$\left[\underline{\underline{M}}\right]^{\gamma} = \cos\left(\alpha_{\gamma} d_{\gamma}\right)\left[\underline{\underline{I}}\right]$$

$$+ \frac{\sin\left(\alpha_{\gamma} d_{\gamma}\right)}{\alpha_{\gamma}} \begin{bmatrix} 0 & 0 & 0 & -\dfrac{i\left(q^2 - \omega^2\varepsilon^{\gamma}\mu_0\right)}{\omega\varepsilon^{\gamma}} \\ 0 & 0 & -i\omega\mu_0 & 0 \\ 0 & \dfrac{i\left(q^2 - \omega^2\varepsilon^{\gamma}\mu_0\right)}{\omega\mu_0} & 0 & 0 \\ i\omega\varepsilon^{\gamma} & 0 & 0 & 0 \end{bmatrix},$$

$$\gamma \in \{\mathcal{A}, \mathcal{B}, \mathcal{C}\}, \quad (5.3)$$

where

$$\alpha_{\gamma} = \sqrt{\omega^2\varepsilon^{\gamma}\mu_0 - q^2}. \quad (5.4)$$

The amplitudes of the transmitted plane wave are related to the amplitudes of the incident and reflected plane waves via the matrix product

$$\left[\underline{\underline{K}}\right]^{-1}_{\text{tr}} \cdot \left[\underline{\underline{M}}\right]^{\mathcal{C}} \cdot \left[\underline{\underline{M}}\right]^{\mathcal{B}} \cdot \left[\underline{\underline{M}}\right]^{\mathcal{A}} \cdot \left[\underline{\underline{K}}\right]_{\text{inc}},$$

according to (3.19). Even when the materials in layers $\mathcal{A}$, $\mathcal{B}$, and $\mathcal{C}$ are isotropic dielectric, this matrix product is too cumbersome to write out explicitly. However, its general form is represented as

$$\left[\underline{\underline{K}}\right]^{-1}_{\text{tr}} \cdot \left[\underline{\underline{M}}\right]^{\mathcal{C}} \cdot \left[\underline{\underline{M}}\right]^{\mathcal{B}} \cdot \left[\underline{\underline{M}}\right]^{\mathcal{A}} \cdot \left[\underline{\underline{K}}\right]_{\text{inc}} = \begin{bmatrix} m_1 & 0 & m_2 & 0 \\ 0 & m_3 & 0 & m_4 \\ m_5 & 0 & m_6 & 0 \\ 0 & m_7 & 0 & m_8 \end{bmatrix}, \quad (5.5)$$

with the entries m_j, $j \in [1, 8]$, being nonzero in general. An immediate consequence of the symmetry of the matrix product represented in (5.5) is that r_p and t_p are independent of a_s, and likewise r_s and t_s are independent of a_p. Therefore, all cross-polarized linear reflection and transmission coefficients are null valued.

In the degenerate case in which $\varepsilon^{\mathcal{A}} = \varepsilon^{\mathcal{B}} = \varepsilon^{\mathcal{C}}$, (3.19) delivers the standard results for the co-polarized linear reflection and transmission coefficients, namely,

$$
\left.
\begin{aligned}
r_{\text{ss}} &= \frac{r_{\text{a}} - r_{\text{b}}}{r_{\text{a}} + r_{\text{b}}} \\[4pt]
r_{\text{pp}} &= \frac{r_{\text{c}} - r_{\text{d}}}{r_{\text{c}} + r_{\text{d}}} \\[4pt]
t_{\text{ss}} &= \frac{2\omega\eta_0\mu_0\alpha_{\mathcal{A}}n_1 \cos\theta_{\text{inc}} \sec\theta_{\text{tr}}}{t_{\text{a}} + t_{\text{b}}} \\[4pt]
t_{\text{pp}} &= \frac{2\omega\eta_0\varepsilon^{\mathcal{A}}\alpha_{\mathcal{A}}n_1 \cos\theta_{\text{inc}} \sec\theta_{\text{tr}}}{t_{\text{c}} + t_{\text{d}}}
\end{aligned}
\right\},
\tag{5.6}
$$

where the scalar parameters

$$
\left.
\begin{aligned}
r_{\text{a}} &= \omega\mu_0 n_1 \cos\theta_{\text{inc}} \left[\omega\mu_0 n_2 \sin(\alpha_{\mathcal{A}}d) + i\alpha_{\mathcal{A}}\eta_0 \cos(\alpha_{\mathcal{A}}d) \sec\theta_{\text{tr}}\right] \\
r_{\text{b}} &= \alpha_{\mathcal{A}}\eta_0 \left[\alpha_{\mathcal{A}}\eta_0 \sec\theta_{\text{tr}} \sin(\alpha_{\mathcal{A}}d) + i\omega\mu_0 n_2 \cos(\alpha_{\mathcal{A}}d)\right] \\
r_{\text{c}} &= \omega\eta_0\varepsilon^{\mathcal{A}} \cos\theta_{\text{inc}} \left[\omega\eta_0\varepsilon^{\mathcal{A}} \sin(\alpha_{\mathcal{A}}d) + i\alpha_{\mathcal{A}}n_2 \sec\theta_{\text{tr}} \cos(\alpha_{\mathcal{A}}d)\right] \\
r_{\text{d}} &= \alpha_{\mathcal{A}}n_1 \left[\alpha_{\mathcal{A}}n_2 \sec\theta_{\text{tr}} \sin(\alpha_{\mathcal{A}}d) + i\omega\eta_0\varepsilon^{\mathcal{A}} \cos(\alpha_{\mathcal{A}}d)\right] \\
t_{\text{a}} &= \eta_0\alpha_{\mathcal{A}} \left[\omega\mu_0 n_2 \cos(\alpha_{\mathcal{A}}d) - i\eta_0\alpha_{\mathcal{A}} \sec\theta_{\text{tr}} \sin(\alpha_{\mathcal{A}}d)\right] \\
t_{\text{b}} &= \omega\mu_0 n_1 \cos\theta_{\text{inc}} \left[\eta_0\alpha_{\mathcal{A}} \sec\theta_{\text{tr}} \cos(\alpha_{\mathcal{A}}d) - i\omega\mu_0 n_2 \sin(\alpha_{\mathcal{A}}d)\right] \\
t_{\text{c}} &= \omega\eta_0\varepsilon^{\mathcal{A}} \cos\theta_{\text{inc}} \left[\alpha_{\mathcal{A}}n_2 \sec\theta_{\text{tr}} \cos(\alpha_{\mathcal{A}}d) - i\omega\eta_0\varepsilon^{\mathcal{A}} \sin(\alpha_{\mathcal{A}}d)\right] \\
t_{\text{d}} &= \alpha_{\mathcal{A}}n_1 \left[\omega\eta_0\varepsilon^{\mathcal{A}} \cos(\alpha_{\mathcal{A}}d) - i\alpha_{\mathcal{A}}n_2 \sec\theta_{\text{tr}} \sin(\alpha_{\mathcal{A}}d)\right]
\end{aligned}
\right\}.
\tag{5.7}
$$

5.2   MULTILAYERED SLAB WITH A SINGLY PERIODIC INTERFACE

Consider the reflection-transmission problem schematically illustrated in Fig. 4.1. A plane wave is incident on a multilayered slab similar to that considered in Section 5.1 (and schematically illustrated in Fig. 3.1) except that the interface of layers $\mathcal{A}$ and $\mathcal{B}$ is periodically corrugated along the x axis. The plane $z = 0$ forms one face of layer $\mathcal{A}$. The interface of layers $\mathcal{A}$ and $\mathcal{B}$ is $z = d_{\mathcal{A}} + g(x)$, where the grating function

$$
g(x) = g(x \pm L_x)
\tag{5.8}
$$

of period L_x satisfies the constraints

$$
\left.
\begin{aligned}
g(x) &< d_{\mathcal{B}} \\
g(x) &> -d_{\mathcal{A}}
\end{aligned}
\right\}, \quad \forall x \in (-\infty, \infty).
\tag{5.9}
$$

The interface of layers $\mathcal{B}$ and $\mathcal{C}$ is the plane $z = d_A + d_B$. Finally, the other face of layer $\mathcal{C}$ is the plane $z = d$ with $d = d_A + d_B + d_C$.

Each of the layers $\mathcal{A}$, $\mathcal{B}$, and $\mathcal{C}$ is made of a different homogeneous isotropic dielectric material. The corresponding constitutive relations of the materials occupying the region $0 < z < d$ are

$$
\underline{D}(\underline{r}) = \left\{
\begin{array}{ll}
\varepsilon^A \underline{E}(\underline{r}), & 0 < z < d_A + g(x) \\
\varepsilon^B \underline{E}(\underline{r}), & d_A + g(x) < z < d_A + d_B \\
\varepsilon^C \underline{E}(\underline{r}), & d_A + d_B < z < d
\end{array}
\right\},
$$
$$
\underline{B}(\underline{r}) = \mu_0 \underline{H}(\underline{r}), \qquad 0 < z < d
$$
(5.10)

with ε^A, ε^B, and ε^C being the permittivities of the materials of the layers $\mathcal{A}$, $\mathcal{B}$, and $\mathcal{C}$, respectively. For simplicity, the half-spaces $z < 0$ and $z > d$ are vacuous, i.e., $n_1 = n_2 = 1$.

5.2.1 INCIDENT PLANE WAVE

In the vacuous half-space $z < 0$, suppose that a plane wave propagating in the xz plane at an angle $\theta_{\text{inc}} \in [0°, 90°)$ to the z axis is incident on the plane $z = 0$. Thus, the plane of propagation of the incident plane wave coincides with the grating plane. While the polarization state of this incident plane wave is arbitrary, the associated field phasors may be decomposed as s- and p-polarized components.

In consideration of the periodically corrugated interface of layers $\mathcal{A}$ and $\mathcal{B}$, the incident field phasors are best expressed as Fourier series with respect to x as follows:

$$
\left.
\begin{array}{l}
\underline{E}_{\text{inc}}(\underline{r}) = \sum_{m \in \mathbb{Z}} \left(a_s^{(m)} \underline{s}^{(m)} + a_p^{(m)} \underline{p}_{\text{inc}}^{(m)} \right) \exp\left[i \left(\kappa^{(m)} x + \alpha^{(m)} z \right) \right] \\
\underline{H}_{\text{inc}}(\underline{r}) = \eta_0^{-1} \sum_{m \in \mathbb{Z}} \left(a_s^{(m)} \underline{p}_{\text{inc}}^{(m)} - a_p^{(m)} \underline{s}^{(m)} \right) \exp\left[i \left(\kappa^{(m)} x + \alpha^{(m)} z \right) \right]
\end{array}
\right\}, \quad z < 0, \quad (5.11)
$$

where the set $\mathbb{Z} = \{0, \pm 1, \pm 2, \dots\}$. The amplitudes in (5.11) are given as

$$
\left.
\begin{array}{l}
a_s^{(m)} = a_s^{(0)} \delta_{m0} \\
a_p^{(m)} = a_p^{(0)} \delta_{m0}
\end{array}
\right\},
$$
(5.12)

using the Kronecker delta $\delta_{mm'}$ defined in (4.10), and the other quantities are as follows:

$$
\left.
\begin{array}{l}
\kappa^{(m)} = k_0 \sin \theta_{\text{inc}} + m \left(2\pi / L_x \right) \\
\alpha^{(m)} = +\sqrt{k_0^2 - \left(\kappa^{(m)} \right)^2} \\
\underline{s}^{(m)} = \underline{u}_y \\
\underline{p}_{\text{inc}}^{(m)} = \dfrac{-\alpha^{(m)} \underline{u}_x + \kappa^{(m)} \underline{u}_z}{k_0}
\end{array}
\right\}.
$$
(5.13)

Each term in the summations on the right sides of (5.11) represents a Floquet harmonic [3]. Furthermore, these are called *linear* Floquet harmonics because the fields are expressed in terms of s- and p-polarization states. Whereas $a_p^{(0)} = 0$ for an s-polarized incident plane wave, $a_s^{(0)} = 0$ for a p-polarized incident plane wave. Circular Floquet harmonics that correspond to circular polarization states, as introduced in Section 4.5 are unnecessary here because the materials involved are isotropic and achiral [4].

5.2.2   REFLECTED AND TRANSMITTED FIELD PHASORS

The reflected field phasors are represented as sums of linear Floquet harmonics as

$$
\left.
\begin{aligned}
\underline{E}_{\mathrm{ref}}(\underline{r}) &= \sum_{m \in \mathbb{Z}} \left(r_s^{(m)} \underline{s}^{(m)} + r_p^{(n)} \underline{p}_{\mathrm{ref}}^{(m)} \right) \exp\left[i \left(\kappa^{(m)} x - \alpha^{(m)} z \right) \right] \\
\underline{H}_{\mathrm{ref}}(\underline{r}) &= \eta_0^{-1} \sum_{m \in \mathbb{Z}} \left(r_s^{(m)} \underline{p}_{\mathrm{ref}}^{(m)} - r_p^{(m)} \underline{s}^{(m)} \right) \exp\left[i \left(\kappa^{(m)} x - \alpha^{(m)} z \right) \right]
\end{aligned}
\right\}, \quad z < 0, \quad (5.14)
$$

wherein $r_s^{(m)}$ and $r_p^{(m)}$ are unknown reflection amplitudes, and the vector

$$
\underline{p}_{\mathrm{ref}}^{(m)} = \frac{\alpha^{(m)} \underline{u}_x + \kappa^{(m)} \underline{u}_z}{k_0}. \tag{5.15}
$$

In a similar vein, the transmitted field phasors are represented as sums of linear Floquet harmonics as

$$
\left.
\begin{aligned}
\underline{E}_{\mathrm{tr}}(\underline{r}) &= \sum_{m \in \mathbb{Z}} \left(t_s^{(m)} \underline{s}^{(m)} + t_p^{(m)} \underline{p}_{\mathrm{inc}}^{(m)} \right) \exp\left\{ i \left[\kappa^{(m)} x + \alpha^{(m)} (z - d) \right] \right\} \\
\underline{H}_{\mathrm{tr}}(\underline{r}) &= \eta_0^{-1} \sum_{m \in \mathbb{Z}} \left(t_s^{(m)} \underline{p}_{\mathrm{inc}}^{(m)} - t_p^{(m)} \underline{s}^{(m)} \right) \exp\left\{ i \left[\kappa^{(m)} x + \alpha^{(m)} (z - d) \right] \right\}
\end{aligned}
\right\},
$$
$$
z > d, \quad (5.16)
$$

where $t_s^{(m)}$ and $t_p^{(m)}$ are unknown transmission amplitudes. The terms in the sums on the right sides of (5.14) and (5.16) that carry the index $m = 0$ are the specular terms. The terms with indexes $m \neq 0$ are the nonspecular terms arising from the periodic nature of the grating function $g(x)$.

The scalar quantity $\alpha^{(m)}$ is positive real when $k_0 > \kappa^{(m)}$, in which case the corresponding Floquet harmonics in (5.14) and (5.16) can transport energy far away from the multilayered slab. Alternatively, $\alpha^{(m)}$ is positive imaginary when $k_0 < \kappa^{(m)}$, in which case the corresponding Floquet harmonics are evanescent.

5.2.3   LINEAR REFLECTANCES AND TRANSMITTANCES

As the direction of propagation of the incident plane wave lies in the grating plane, no depolarization can occur on reflection and transmission. It follows that the cross-polarized reflection and transmission coefficients are null valued. Hence,

$$\left.\begin{aligned}
r_{\mathrm{s}}^{(m)} &= r_{\mathrm{ss}}^{(m)}\, a_{\mathrm{s}}^{(0)} \\
t_{\mathrm{s}}^{(m)} &= t_{\mathrm{ss}}^{(m)}\, a_{\mathrm{s}}^{(0)} \\
r_{\mathrm{p}}^{(m)} &= r_{\mathrm{pp}}^{(m)}\, a_{\mathrm{p}}^{(0)} \\
t_{\mathrm{p}}^{(m)} &= t_{\mathrm{pp}}^{(m)}\, a_{\mathrm{p}}^{(0)}
\end{aligned}\right\}, \tag{5.17}$$

where $r_{\mathrm{ss}}^{(m)}$ and $r_{\mathrm{pp}}^{(m)}$ are the co-polarized reflection coefficients, and $t_{\mathrm{ss}}^{(m)}$ and $t_{\mathrm{pp}}^{(m)}$ are the co-polarized transmission coefficients of order $m \in (-\infty, \infty)$. The corresponding linear reflectances and transmittances are defined as

$$\left.\begin{aligned}
R_{\mathrm{ss}}^{(m)} &= \frac{\mathrm{Re}\left\{\alpha^{(m)}\right\}}{\alpha^{(0)}}\, |r_{\mathrm{ss}}^{(m)}|^2 \\
T_{\mathrm{ss}}^{(m)} &= \frac{\mathrm{Re}\left\{\alpha^{(m)}\right\}}{\alpha^{(0)}}\, |t_{\mathrm{ss}}^{(m)}|^2 \\
R_{\mathrm{pp}}^{(m)} &= \frac{\mathrm{Re}\left\{\alpha^{(m)}\right\}}{\alpha^{(0)}}\, |r_{\mathrm{pp}}^{(m)}|^2 \\
T_{\mathrm{pp}}^{(m)} &= \frac{\mathrm{Re}\left\{\alpha^{(m)}\right\}}{\alpha^{(0)}}\, |t_{\mathrm{pp}}^{(m)}|^2
\end{aligned}\right\}. \tag{5.18}$$

The real-valued quantities

$$\left.\begin{aligned}
A_{\mathrm{s}} &= 1 - \sum_{m \in \mathbb{Z}} \left(R_{\mathrm{ss}}^{(m)} + T_{\mathrm{ss}}^{(m)} \right) \\
A_{\mathrm{p}} &= 1 - \sum_{m \in \mathbb{Z}} \left(R_{\mathrm{pp}}^{(m)} + T_{\mathrm{pp}}^{(m)} \right)
\end{aligned}\right\} \tag{5.19}$$

are the absorptances for s- and p-polarized incident plane waves, respectively. Provided that the materials of the layers $\mathcal{A}$, $\mathcal{B}$, and $\mathcal{C}$ are passive, neither of the two linear absorptances A_{s} and A_{p} can be negative. Also, the principle of conservation of energy guarantees that A_{s} and A_{p} do not exceed unity.

5.2.4 RIGOROUS COUPLED-WAVE APPROACH

The RCWA [5–7] is very well suited and is widely used to calculate the amplitudes of the Floquet harmonics in the reflected and transmitted field phasors presented in Section 5.2.2. First, the constitutive relations of the materials of layers $\mathcal{A}$, $\mathcal{B}$, and $\mathcal{C}$ are jointly written as

$$\left.\begin{aligned}
\underline{D}(r) &= \varepsilon(x, z)\, \underline{E}(r) \\
\underline{B}(r) &= \mu_0\, \underline{H}(r)
\end{aligned}\right\}, \quad z \in (0, d), \tag{5.20}$$

where

$$
\varepsilon(x,z) = \begin{cases}
\varepsilon^A , & z \in (0, d_A + g_{\min}] , \\
\varepsilon^A + (\varepsilon_B - \varepsilon_A)\,\mathcal{U}\,[z - d_A - g(x)] , & z \in (d_A + g_{\min}, d_A + g_{\max}) , \\
\varepsilon^B , & z \in [d_A + g_{\max}, d_A + d_B) , \\
\varepsilon^C , & z \in [d_A + d_B, d) ,
\end{cases}
\tag{5.21}
$$

with the unit step function $\mathcal{U}(s)$ defined in (4.29). Thereafter, the Fourier series

$$
\varepsilon(x,z) = \sum_{m\in\mathbb{Z}} \varepsilon^{(m)}(z) \exp\left(i 2\pi m x / L_x\right) , \quad z \in (0,d) ,
\tag{5.22}
$$

is obtained with coefficients

$$
\varepsilon^{(m)}(z) = \begin{cases}
\varepsilon^A \delta_{m0}, & z \in (0, d_A + g_{\min}] \\
\varepsilon^A \delta_{m0} + \left(\varepsilon^B - \varepsilon^A\right) \Upsilon^{(m)}(z), & z \in (d_A + g_{\min}, d_A + g_{\max}) \\
\varepsilon^B \delta_{m0}, & z \in [d_A + g_{\max}, d_A + d_B) \\
\varepsilon^C \delta_{m0}, & z \in [d_A + d_B, d)
\end{cases} ,
\tag{5.23}
$$

where

$$
\Upsilon^{(m)}(z) = \frac{1}{L_x} \int_0^{L_x} \mathcal{U}\,[z - d_A - g(x)] \exp\left(-i 2\pi m x / L_x\right) dx .
\tag{5.24}
$$

Next, the variations of the field phasors inside the region $0 < z < d$ with respect to x are represented as the Fourier series

$$
\left.
\begin{aligned}
\underline{E}(\underline{r}) &= \sum_{m\in\mathbb{Z}} \left[\underline{e}^{(m)}(z) \exp\left(i\kappa^{(m)} x\right)\right] \\
\underline{H}(\underline{r}) &= \sum_{m\in\mathbb{Z}} \left[\underline{h}^{(m)}(z) \exp\left(i\kappa^{(m)} x\right)\right]
\end{aligned}
\right\} , \quad z \in (0,d) .
\tag{5.25}
$$

The vector functions

$$
\left.
\begin{aligned}
\underline{e}^{(m)}(z) &= e_x^{(m)}\underline{u}_x + e_y^{(m)}\underline{u}_y + e_z^{(m)}\underline{u}_z \\
\underline{h}^{(m)}(z) &= h_x^{(m)}\underline{u}_x + h_y^{(m)}\underline{u}_y + h_z^{(m)}\underline{u}_z
\end{aligned}
\right\}
\tag{5.26}
$$

are to be determined for all m.

The combination of (5.20), (5.22), and (5.25) with the source-free frequency-domain Maxwell curl postulates

$$
\left.
\begin{aligned}
\nabla \times \underline{E}(\underline{r}) &= i\omega\mu_0 \underline{H}(\underline{r}) \\
\nabla \times \underline{H}(\underline{r}) &= -i\omega\varepsilon(x,z)\,\underline{E}(\underline{r})
\end{aligned}
\right\} , \quad z \in (0,d),
\tag{5.27}
$$

yields the systems of four ordinary differential equations

$$
\left.
\begin{aligned}
&\frac{d}{dz}e_x^{(m)}(z) - i\kappa^{(m)}e_z^{(m)} = i\omega\mu_0\,h_y^{(m)}(z) \\
&\frac{d}{dz}e_y^{(m)}(z) = -i\omega\mu_0\,h_x^{(m)}(z) \\
&\frac{d}{dz}h_x^{(m)}(z) - i\kappa^{(m)}h_z^{(n)} = -i\omega\sum_{m'\in\mathbb{Z}}\left[\varepsilon^{(m-m')}(z)\,e_y^{(m')}(z)\right] \\
&\frac{d}{dz}h_y^{(m)}(z) = i\omega\sum_{m'\in\mathbb{Z}}\left[\varepsilon^{(m-m')}(z)\,e_x^{(m')}(z)\right]
\end{aligned}
\right\}\,,\quad m\in(-\infty,\infty)\,,\quad (5.28)
$$

and two algebraic equations

$$
\left.
\begin{aligned}
&\kappa^{(m)}e_y^{(m)} = \omega\mu_0\,h_z^{(m)}(z) \\
&\kappa^{(m)}h_y^{(m)}(z) = -\omega\sum_{m'\in\mathbb{Z}}\left[\varepsilon^{(m-m')}(z)\,e_z^{(m')}(z)\right]
\end{aligned}
\right\}\,,\quad m\in(-\infty,\infty)\,,\quad (5.29)
$$

for all $z\in(0,d)$.

Equations (5.28) and (5.29) involve sums containing an infinite number of terms and are themselves infinite in number. Truncation is required for numerical tractability. The index m in (5.25) is restricted to the range $[-M_t, M_t]$. For use in the region $0 < z < d$, the $(2M_t + 1)$-column vectors[1] $[\breve{e}_\nu(z)]$ and $\left[\breve{h}_\nu(z)\right]$ are introduced for $\nu\in\{x,y,z\}$, as defined in (4.37) but with $N_t = 0$. Introduction of the $(2M_t + 1) \times (2M_t + 1)$ matrixes

$$
\left[\underline{\underline{\breve{K}}}\right] = \mathrm{diag}\left[\kappa^{(-M_t)},\ \kappa^{(-M_t+1)},\ ...,\ \kappa^{(M_t-1)},\ \kappa^{(M_t)}\right]\,,\quad (5.30)
$$

and

$$
\left[\underline{\underline{\breve{\varepsilon}}(z)}\right] =
$$

$$
\begin{bmatrix}
\varepsilon^{(0)}(z) & \varepsilon^{(-1)}(z) & \varepsilon^{(-2)}(z) & \cdots & \varepsilon^{(-2M_t+1)}(z) & \varepsilon^{(-2M_t)}(z) \\
\varepsilon^{(1)}(z) & \varepsilon^{(0)}(z) & \varepsilon^{(-1)}(z) & \cdots & \varepsilon^{(-2M_t+2)}(z) & \varepsilon^{(-2M_t+1)}(z) \\
\cdots & \cdots & \cdots & \cdots & \cdots & \cdots \\
\varepsilon^{(2M_t-1)}(z) & \varepsilon^{(2M_t-2)}(z) & \varepsilon^{(2M_t-3)}(z) & \cdots & \varepsilon^{(0)}(z) & \varepsilon^{(-1)}(z) \\
\varepsilon^{(2M_t)}(z) & \varepsilon^{(2M_t-1)}(z) & \varepsilon^{(2M_t-2)}(z) & \cdots & \varepsilon^{(1)}(z) & \varepsilon^{(0)}(z)
\end{bmatrix}\,,\quad (5.31)
$$

in (5.29) delivers the relations

$$
\left.
\begin{aligned}
&[\breve{e}_z(z)] = -\omega^{-1}\left[\underline{\underline{\breve{\varepsilon}}(z)}\right]^{-1}\cdot\left[\underline{\underline{\breve{K}}}\right]\cdot\left[\breve{h}_y(z)\right] \\
&\left[\breve{h}_z(z)\right] = (\omega\mu_0)^{-1}\left[\underline{\underline{\breve{K}}}\right]\cdot[\breve{e}_y(z)]
\end{aligned}
\right\}\,.\quad (5.32)
$$

[1]The symbol ˘ identifies quantities associated with the RCWA.

Equations (5.32) allow $e_z^{(m)}(z)$ and $h_z^{(m)}(z)$ to be eliminated from (5.28). Hence, the matrix ordinary differential equation

$$\frac{d}{dz}\left[\underline{\breve{f}}(z)\right] = i\left[\underline{\underline{\breve{P}}}(z)\right] \cdot \left[\underline{\breve{f}}(z)\right], \quad z \in (0, d),\tag{5.33}$$

emerges, wherein the $4(2M_t + 1)$-column vector

$$\left[\underline{\breve{f}}(z)\right] = \begin{bmatrix} [\breve{e}_x(z)] \\ [\breve{e}_y(z)] \\ [\breve{h}_x(z)] \\ [\breve{h}_y(z)] \end{bmatrix}\tag{5.34}$$

and the $4(2M_t + 1) \times 4(2M_t + 1)$ matrix

$$\left[\underline{\underline{\breve{P}}}(z)\right] = \omega \begin{bmatrix} \left[\underline{\underline{\breve{0}}}\right] & \left[\underline{\underline{\breve{0}}}\right] & \left[\underline{\underline{\breve{0}}}\right] & \mu_0\left[\underline{\underline{\breve{I}}}\right] \\ \left[\underline{\underline{\breve{0}}}\right] & \left[\underline{\underline{\breve{0}}}\right] & -\mu_0\left[\underline{\underline{\breve{I}}}\right] & \left[\underline{\underline{\breve{0}}}\right] \\ \left[\underline{\underline{\breve{0}}}\right] & -\left[\underline{\underline{\breve{\varepsilon}}}(z)\right] & \left[\underline{\underline{\breve{0}}}\right] & \left[\underline{\underline{\breve{0}}}\right] \\ \left[\underline{\underline{\breve{\varepsilon}}}(z)\right] & \left[\underline{\underline{\breve{0}}}\right] & \left[\underline{\underline{\breve{0}}}\right] & \left[\underline{\underline{\breve{0}}}\right] \end{bmatrix}$$
$$+ \frac{1}{\omega}\begin{bmatrix} \left[\underline{\underline{\breve{0}}}\right] & \left[\underline{\underline{\breve{0}}}\right] & \left[\underline{\underline{\breve{0}}}\right] & -\left[\underline{\underline{\breve{K}}}\right] \cdot \left[\underline{\underline{\breve{\varepsilon}}}(z)\right]^{-1} \cdot \left[\underline{\underline{\breve{K}}}\right] \\ \left[\underline{\underline{\breve{0}}}\right] & \left[\underline{\underline{\breve{0}}}\right] & \left[\underline{\underline{\breve{0}}}\right] & \left[\underline{\underline{\breve{0}}}\right] \\ \left[\underline{\underline{\breve{0}}}\right] & \mu_0^{-1}\left[\underline{\underline{\breve{K}}}\right] \cdot \left[\underline{\underline{\breve{K}}}\right] & \left[\underline{\underline{\breve{0}}}\right] & \left[\underline{\underline{\breve{0}}}\right] \\ \left[\underline{\underline{\breve{0}}}\right] & \left[\underline{\underline{\breve{0}}}\right] & \left[\underline{\underline{\breve{0}}}\right] & \left[\underline{\underline{\breve{0}}}\right] \end{bmatrix}.\tag{5.35}$$

Each submatrix in the definition of $\left[\underline{\underline{\breve{P}}}(z)\right]$ has dimension $(2M_t + 1) \times (2M_t + 1)$, $\left[\underline{\underline{\breve{0}}}\right]$ is the $(2M_t + 1) \times (2M_t + 1)$ null matrix, and $\left[\underline{\underline{\breve{I}}}\right]$ is the $(2M_t + 1) \times (2M_t + 1)$ identity matrix.

In order to solve (5.33), boundary conditions on $\left[\underline{\breve{f}}(z)\right]$ are implemented at $z = 0$ and $z = d$. The required boundary values $\left[\underline{\breve{f}}(0-)\right]$ and $\left[\underline{\breve{f}}(d+)\right]$ are derived from (5.11), (5.14), and (5.16), as follows: The amplitudes of the s-polarized and p-polarized components of the incident, reflected, and transmitted phasors are assembled into $2(2M_t + 1)$-column vectors

$$
\left.
\begin{aligned}
\left[\underline{\breve{A}}\right] &= [a_{\mathrm{s}}^{(-M_t)}, a_{\mathrm{s}}^{(-M_t+1)}, \ldots, a_{\mathrm{s}}^{(M_t-1)}, a_{\mathrm{s}}^{(M_t)}, a_{\mathrm{p}}^{(-M_t)}, a_{\mathrm{p}}^{(-M_t+1)}, \ldots, a_{\mathrm{p}}^{(M_t-1)}, a_{\mathrm{p}}^{(M_t)}]^T \\
\left[\underline{\breve{R}}\right] &= [r_{\mathrm{s}}^{(-M_t)}, r_{\mathrm{s}}^{(-M_t+1)}, \ldots, r_{\mathrm{s}}^{(M_t-1)}, r_{\mathrm{s}}^{(M_t)}, r_{\mathrm{p}}^{(-M_t)}, r_{\mathrm{p}}^{(-M_t+1)}, \ldots, r_{\mathrm{p}}^{(M_t-1)}, r_{\mathrm{p}}^{(M_t)}]^T \\
\left[\underline{\breve{T}}\right] &= [t_{\mathrm{s}}^{(-M_t)}, t_{\mathrm{s}}^{(-M_t+1)}, \ldots, t_{\mathrm{s}}^{(M_t-1)}, t_{\mathrm{s}}^{(M_t)}, t_{\mathrm{p}}^{(-M_t)}, t_{\mathrm{p}}^{(-M_t+1)}, \ldots, t_{\mathrm{p}}^{(M_t-1)}, t_{\mathrm{p}}^{(M_t)}]^T
\end{aligned}
\right\}.
$$

(5.36)

The boundary values are then expressed as

$$
\left[\underline{\breve{f}}(0-)\right] =
\begin{bmatrix}
\left[\underline{\breve{0}}\right] & -k_0^{-1}\left[\underline{\underline{\breve{\alpha}}}\right] & \left[\underline{\breve{0}}\right] & k_0^{-1}\left[\underline{\underline{\breve{\alpha}}}\right] \\
\left[\underline{\underline{\breve{I}}}\right] & \left[\underline{\breve{0}}\right] & \left[\underline{\underline{\breve{I}}}\right] & \left[\underline{\breve{0}}\right] \\
-\eta_0^{-1}k_0^{-1}\left[\underline{\underline{\breve{\alpha}}}\right] & \left[\underline{\breve{0}}\right] & \eta_0^{-1}k_0^{-1}\left[\underline{\underline{\breve{\alpha}}}\right] & \left[\underline{\breve{0}}\right] \\
\left[\underline{\breve{0}}\right] & -\eta_0^{-1}\left[\underline{\underline{\breve{I}}}\right] & \left[\underline{\breve{0}}\right] & -\eta_0^{-1}\left[\underline{\underline{\breve{I}}}\right]
\end{bmatrix}
\cdot
\begin{bmatrix}
\left[\underline{\breve{A}}\right] \\
\left[\underline{\breve{R}}\right]
\end{bmatrix}
$$

(5.37)

and

$$
\left[\underline{\breve{f}}(d+)\right] =
\begin{bmatrix}
\left[\underline{\breve{0}}\right] & -k_0^{-1}\left[\underline{\underline{\breve{\alpha}}}\right] \\
\left[\underline{\underline{\breve{I}}}\right] & \left[\underline{\breve{0}}\right] \\
-\eta_0^{-1}k_0^{-1}\left[\underline{\underline{\breve{\alpha}}}\right] & \left[\underline{\breve{0}}\right] \\
\left[\underline{\breve{0}}\right] & -\eta_0^{-1}\left[\underline{\underline{\breve{I}}}\right]
\end{bmatrix}
\cdot
\left[\underline{\breve{T}}\right],
$$

(5.38)

wherein the $(2M_t + 1) \times (2M_t + 1)$ diagonal matrix

$$
\left[\underline{\underline{\breve{\alpha}}}\right] = \mathrm{diag}\left[\alpha^{(-M_t)}, \alpha^{(-M_t+1)}, \ldots, \alpha^{(M_t-1)}, \alpha^{(M_t)}\right].
$$

(5.39)

The boundary values $\left[\underline{\breve{f}}(0-)\right]$ and $\left[\underline{\breve{f}}(d+)\right]$ are more conveniently recast as

$$
\left[\underline{\breve{f}}(0-)\right] =
\begin{bmatrix}
\left[\underline{\underline{\breve{Y}}}_e^{\mathrm{inc}}\right] & \left[\underline{\underline{\breve{Y}}}_e^{\mathrm{ref}}\right] \\
\left[\underline{\underline{\breve{Y}}}_h^{\mathrm{inc}}\right] & \left[\underline{\underline{\breve{Y}}}_h^{\mathrm{ref}}\right]
\end{bmatrix}
\cdot
\begin{bmatrix}
\left[\underline{\breve{A}}\right] \\
\left[\underline{\breve{R}}\right]
\end{bmatrix},
$$

(5.40)

and

$$
\left[\underline{\breve{f}}(d)\right] =
\begin{bmatrix}
\left[\underline{\underline{\breve{Y}}}_e^{\mathrm{tr}}\right] \\
\left[\underline{\underline{\breve{Y}}}_h^{\mathrm{tr}}\right]
\end{bmatrix}
\cdot
\left[\underline{\breve{T}}\right].
$$

(5.41)

The $2(2M_t + 1) \times 2(2M_t + 1)$ matrixes on the right side of (5.40) are extracted from (5.37) by direct inspection as

$$\left[\underline{\underline{\breve{Y}}}_e^{\text{inc}}\right] = \begin{bmatrix} \left[\underline{\underline{\breve{0}}}\right] & -k_0^{-1}\left[\underline{\underline{\breve{\alpha}}}\right] \\ \left[\underline{\underline{\breve{I}}}\right] & \left[\underline{\underline{\breve{0}}}\right] \end{bmatrix}, \quad \left[\underline{\underline{\breve{Y}}}_h^{\text{inc}}\right] = -\eta_0^{-1}\begin{bmatrix} k_0^{-1}\left[\underline{\underline{\breve{\alpha}}}\right] & \left[\underline{\underline{\breve{0}}}\right] \\ \left[\underline{\underline{\breve{0}}}\right] & \left[\underline{\underline{\breve{I}}}\right] \end{bmatrix}$$

$$\left[\underline{\underline{\breve{Y}}}_e^{\text{ref}}\right] = \begin{bmatrix} \left[\underline{\underline{\breve{0}}}\right] & k_0^{-1}\left[\underline{\underline{\breve{\alpha}}}\right] \\ \left[\underline{\underline{\breve{I}}}\right] & \left[\underline{\underline{\breve{0}}}\right] \end{bmatrix}, \quad \left[\underline{\underline{\breve{Y}}}_h^{\text{ref}}\right] = \eta_0^{-1}\begin{bmatrix} k_0^{-1}\left[\underline{\underline{\breve{\alpha}}}\right] & \left[\underline{\underline{\breve{0}}}\right] \\ \left[\underline{\underline{\breve{0}}}\right] & -\left[\underline{\underline{\breve{I}}}\right] \end{bmatrix}$$

$$\qquad (5.42)$$

Also,

$$\left.\begin{aligned} \left[\underline{\underline{\breve{Y}}}_e^{\text{tr}}\right] &= \left[\underline{\underline{\breve{Y}}}_e^{\text{inc}}\right] \\ \left[\underline{\underline{\breve{Y}}}_h^{\text{tr}}\right] &= \left[\underline{\underline{\breve{Y}}}_h^{\text{inc}}\right] \end{aligned}\right\}, \qquad (5.43)$$

since both half-spaces $z > d$ and $z < 0$ are occupied by the same medium (i.e., vacuum).

The solution of (5.33) can now be obtained in the same way as of (4.47) in Section 4.6. The stable algorithm described in Section 4.7 should be used to circumvent computational difficulties [8, 9]. A few other numerical strategies have emerged for isotropic dielectric materials [10–13], but their efficacies need investigation on a case-by-case basis [14, 15].

5.3   SAMPLE NUMERICAL RESULTS

Plots of linear reflectances, transmittances, and absorptances vs. angle of incidence θ_{inc} are provided in Figs. 5.1–5.3, for representative multilayered slabs comprising layers that are made of isotropic dielectric materials. That is, each of the layers $\mathcal{A}$, $\mathcal{B}$, and $\mathcal{C}$ is made of an isotropic dielectric material, as specified by the permittivities $\varepsilon^\mathcal{A}$, $\varepsilon^\mathcal{B}$, and $\varepsilon^\mathcal{C}$. The parameters used for the calculations are specified in the captions of Figs. 5.1–5.3. The Mathematica™ code used to generate the numerical data plotted in these figures is provided in Appendix B.

Calculated data are presented in Fig. 5.1 for a multilayered slab with planar interfaces. For both s- and p-polarized incident light, the absorptance is relatively large and remains almost constant as θ_{inc} increases from $0°$, but it approaches zero as θ_{inc} approaches $90°$. Where the incident light is s- or p-polarized, the reflectance is very low for $\theta_{\text{inc}} = 0°$, but it approaches unity as θ_{inc} approaches $90°$; conversely, the transmittance is moderate for $\theta_{\text{inc}} = 0°$, but it approaches zero as θ_{inc} approaches $90°$.

The case of a multilayered slab with a periodic interface at $z = d_\mathcal{A} + g(x)$ is represented in Fig. 5.2. The chosen grating function is

$$g(x) = \frac{L_g}{2}\sin\left(\frac{2\pi x}{L_x}\right). \qquad (5.44)$$

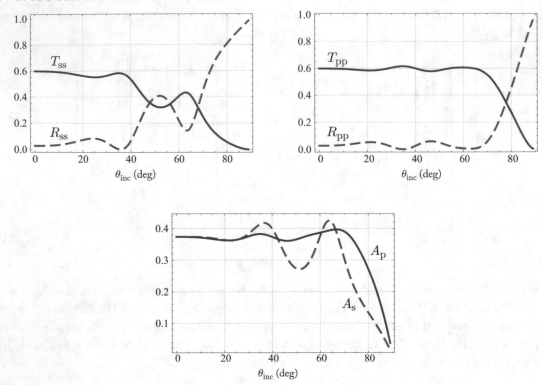

Figure 5.1: Linear reflectances, transmittances, and absorptances plotted against angle of incidence θ_{inc} for an isotropic dielectric multilayered slab with planar interfaces. Permittivity values: $\varepsilon^{\mathcal{A}} = (2.9 + 0.01i)\,\varepsilon_0$, $\varepsilon^{\mathcal{B}} = (4 + 0.03i)\,\varepsilon_0$, and $\varepsilon^{\mathcal{C}} = (3 + 0.05i)\,\varepsilon_0$. Also, $d_{\mathcal{A}} = 0.9\lambda_0$, $d_{\mathcal{B}} = 1.8\lambda_0$, $d_{\mathcal{C}} = 1.2\lambda_0$, with $\lambda_0 = 560$ nm, and $n_1 = n_2 = 1$.

Aside from the periodically corrugated interface with $L_g = \lambda_0/40$, the multilayered slab for Fig. 5.2 is the same as that for Fig. 5.1. The characteristics of the specular reflectances and transmittances, as well as the absorptances, plotted in Fig. 5.2 are qualitatively similar to those of the reflectances, transmittances, and absorptances plotted in Fig. 5.1. The only nonspecular reflectances and transmittances that exist are those labeled by the superscript (-1); and these exist for $6.5° < \theta_{inc} < 90°$. The nonspecular reflectances and transmittances plotted in Fig. 5.2 are much smaller in magnitudes than their specular counterparts. For both s- and p-polarized incident light, the nonspecular reflectances and transmittances approach zero as θ_{inc} approaches $90°$.

The magnitudes of the nonspecular reflectances and transmittances presented in Fig. 5.2 are relatively small because the depth L_g of the corrugations is very small. On increasing L_g, the magnitudes of the nonspecular reflectances and transmittances increase substantially, as can

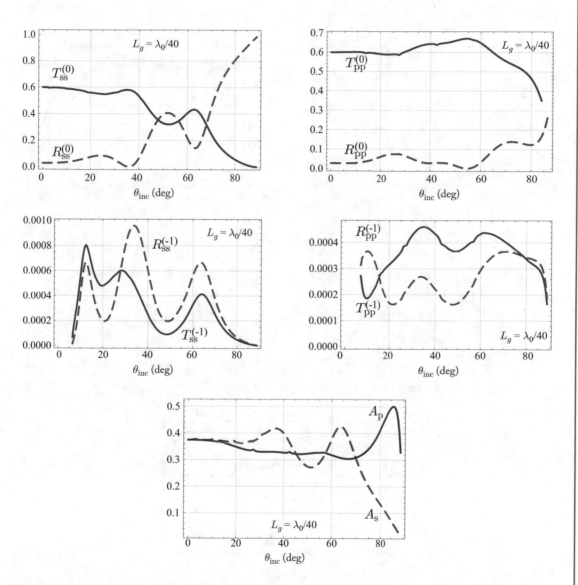

Figure 5.2: As Fig. 5.1 but for an isotropic dielectric multilayered slab with a periodic interface at $z = d_A + g(x)$. The grating function has the sinusoidal form given in (5.44) with $L_g = \lambda_0/40$ and $L_x = 0.9\lambda_0$. Numerical parameters: $M_t = 4$, $N_I = 1$, $N_{II} = 5$, $N_{III} = 1$, and $N_{IV} = 1$.

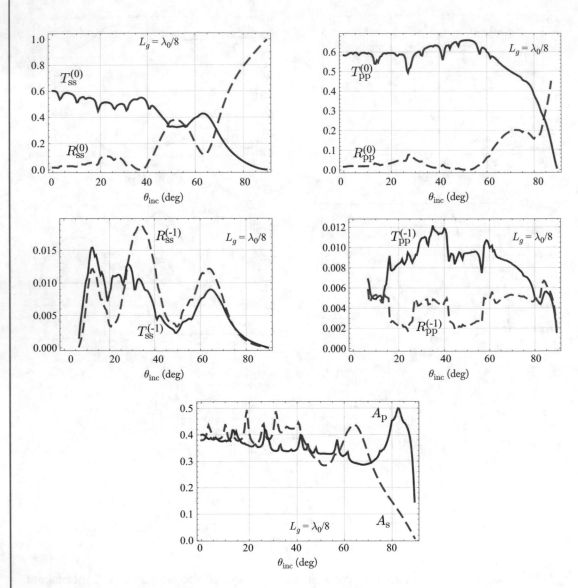

Figure 5.3: As Fig. 5.2, but for $L_g = \lambda_0/8$ with $M_t = 8$ and $N_{II} = 12$.

be appreciated from Fig. 5.3 wherein the reflectances and transmittances, and absorptances are plotted for the same case as Fig. 5.2 except that $L_g = \lambda_0/8$. The specular reflectances and transmittances are largely unaffected by the change in L_g. In order to achieve numerical convergence, a greater number of Floquet harmonics and a greater number of subregions N_{II} are needed when L_g is increased: $M_t = 4$ and $N_{II} = 5$ for Fig. 5.2, whereas $M_t = 8$ and $N_{II} = 12$ for Fig. 5.3.

5.4   REFERENCES

[1] Lakhtakia, A. and Messier, R. 2005. *Sculptured Thin Films: Nanoengineered Morphology and Optics* (Bellingham, WA, SPIE Press). DOI: 10.1117/3.585322. 75

[2] Polo, J. A., Jr., Mackay, T. G., and Lakhtakia, A. 2013. *Electromagnetic Surface Waves: A Modern Perspective* (Waltham, MA, Elsevier). 75

[3] Waterman, P. C. 1975. Scattering by periodic surfaces, *J. Acoust. Soc. Am.*, 57:791–802. DOI: 10.1121/1.380521. 79

[4] Mackay, T. G. and Lakhtakia, A. 2019. *Electromagnetic Anisotropy and Bianisotropy: A Field Guide*, 2nd ed. (Singapore: World Scientific). DOI: 10.1142/11351. 79

[5] Moharam, M. G. and Gaylord, T. K. 1981. Rigorous coupled-wave analysis of planar-grating diffraction, *J. Opt. Soc. Am.*, 71:811–818. DOI: 10.1364/josa.71.000811. 80

[6] Chateau, N. and Hugonin, J.-P. 1994. Algorithm for the rigorous coupled-wave analysis of grating diffraction, *J. Opt. Soc. Am. A*, 11:1321–1331. DOI: 10.1364/josaa.11.001321. 80

[7] Faryad, M. and Lakhtakia, A. 2011. Grating-coupled excitation of multiple surface plasmon-polariton waves. *Phys. Rev. A*, 84:033852. DOI: 10.1103/physreva.84.033852. 80

[8] Li, L. 1993. Multilayer modal method for diffraction gratings of arbitrary profile, depth, and permittivity, *J. Opt. Soc. Am. A*, 10:2581–2591. DOI: 10.1364/JOSAA.10.002581. Li, L. 1994. Multilayer modal method for diffraction gratings of arbitrary profile, depth, and permittivity *J. Opt. Soc. Am. A* 11:1685 (addendum). DOI: 10.1364/JOSAA.11.001685. 85

[9] Wang, F., Horn, M. W., and Lakhtakia, A. 2004. Rigorous electromagnetic modeling of near-field phase-shifting contact lithography, *Microelectron. Eng.*, 71:34–53. DOI: 10.1016/j.mee.2003.09.003. 85

[10] Lalanne, P. and Morris, G. M. 1996. Highly improved convergence of the coupled-wave method for TM polarization, *J. Opt. Soc. Am. A*, 13:779–784. DOI: 10.1364/josaa.13.000779. 85

[11] Granet, G. and Guizal, B. 1996. Efficient implementation of the coupled-wave method for metallic lamellar gratings in TM polarization, *J. Opt. Soc. Am. A*, 13:1019–1023. DOI: 10.1364/josaa.13.001019. 85

[12] Schuster, T., Ruoff, J., Kerwien, N., Rafler, S., and Osten, W. 2007. Normal vector method for convergence improvement using the RCWA for crossed gratings, *J. Opt. Soc. Am. A*, 24:2880–2890. DOI: 10.1364/josaa.24.002880. 85

[13] Auer, M. and Brenner, K.-H. 2014. Localized input fields in rigorous coupled-wave analysis, *J. Opt. Soc. Am. A*, 31:2385–2393. DOI: 10.1364/josaa.31.002385. 85

[14] Lalanne, P. 1997. Convergence performance of the coupled-wave and the differential methods for thin gratings, *J. Opt. Soc. Am. A*, 14:1583–1591. DOI: 10.1364/josaa.14.001583. 85

[15] Anderson, T. H., Civiletti, B. J., Monk, P. B., and Lakhtakia, A. 2020. Coupled optoelectronic simulation and optimization of thin-film photovoltaic solar cells, *J. Comput. Phys.*, 407:109242. DOI: 10.1016/j.jcp.2020.109242. 85

CHAPTER 6

Epilogue

How are the reflected light and the transmitted light related to the light incident on a slab? This fundamental question—which is the quintessential question in thin-film optics [1]—is often tackled by means of the TMM. The slab under consideration may be a single layer of a homogeneous material. Alternatively, the slab may be nonhomogeneous in the thickness direction: the nonhomogeneity may be continuously varying as in the case of certain sculptured thin films [2] and liquid crystals [3] or discontinuously varying in which case the slab is multilayered.

In the preceding three chapters, the TMM is presented for a three-layered slab. The case of a homogeneous slab arises directly a special case, and a straightforward extension can be made for a slab with more than three layers. The formalism also applies to a continuously nonhomogeneous slab, by using the piecewise-uniform approximation [2]. An important feature of the TMM is its accommodation of a multilayered slab whose adjacent constituent layers have periodically corrugated interfaces. This accommodation is achieved by using the rigorous coupled-wave approach [4–9], which involves expansions of field phasors in terms of Floquet harmonics and Fourier expansions of constitutive parameters [10, 11].

Since the TMM is essentially based on an experimental scenario (i.e., a light beam incident on a slab), it has many practical applications. For example, the TMM combined with ellipsometry measurements [12, 13] can be used to characterize the optical properties of the slab material. Thereby, the TMM can aid the design of optical components such as Bragg mirrors and antireflection coatings [14, 15]. A principal application for multilayered slabs is in the analysis of surface waves excited at interface of two dissimilar constituent layers [9, 16]. These surface waves may be excited by prism-coupling techniques [17–20] or by grating-coupling techniques [21, 22] or some combination thereof [23].

Unlike other approaches to the reflection-transmission boundary-value problem, such as iterative methods [24], the TMM has been developed to accommodate multilayered slabs made of the most general linear materials, i.e., bianisotropic materials [25]. Such materials exhibit a rich palette of electromagnetic and optical phenomenons. As an illustration, the planewave dispersion relation for bianisotropic materials can yield up to four independent solutions; in contrast, the corresponding dispersion relations for anisotropic dielectric materials generally yield only two independent solutions and for isotropic dielectric-magnetic materials yield only one independent solution [26]. Consequently, bianisotropic materials can exhibit exotic planewave characteristics, such as propagation with negative phase velocity and negative refraction [27, 28]. While naturally occurring materials that are appreciably bianisotropic are relatively scarce (under normal environmental conditions), engineered bianisotropic materials are readily conceptualized via the process of homogenization [29]. Thus, bianisotropic materials look set to play an increasingly important role as developments in nanotechnology [30, 31] and metamaterial science [32, 33] unfold.

As mentioned in Section 2.5.2, the transfer matrix $\left[\underline{\underline{M}}\right]$ in electromagnetics and optics is a 4×4 matrix that emerges from the frequency-domain Maxwell curl postulates without taking recourse to differentiation with respect to z. Provided that the equations of motion in solid mechanics are also expressed as first-order partial differential equations, the concept of a transfer matrix can be applied for time-harmonic mechanical waves as well, $\left[\underline{\underline{M}}\right]$ then being a 6×6 matrix. Extension to piezoelectric solids leads to $\left[\underline{\underline{M}}\right]$ being a 10×10 matrix [34, 35].

6.1   REFERENCES

[1] Heavens, O. S. 1960. Optical properties of thin films, *Rep. Prog. Phys.*, 23:1–65. DOI: 10.1088/0034-4885/23/1/301. 91

[2] Lakhtakia, A. and Messier, R. 2005. *Sculptured Thin Films: Nanoengineered Morphology and Optics* (Bellingham, WA, SPIE Press). DOI: 10.1117/3.585322. 91

[3] Abdulhalim, I., Benguigui, L., and Weil, R. 1985. Selective reflection by helicoidal liquid crystals. Results of an exact calculation using the 4×4 characteristic matrix method, *J. Phys.*, 46:815–825, Paris. DOI: 10.1051/jphys:01985004605081500. 91

[4] Glytsis, E. N. and Gaylord, T. K. 1987. Rigorous three-dimensional coupled-wave diffraction analysis of single and cascaded anisotropic gratings, *J. Opt. Soc. Am. A*, 4:2061–2080, DOI: 10.1364/josaa.4.002061. 91

[5] Moharam, M. G. and Gaylord, T. K. 1981. Rigorous coupled-wave analysis of planar-grating diffraction, *J. Opt. Soc. Am.*, 71:811–818. DOI: 10.1364/josa.71.000811. 91

[6] Chateau, N. and Hugonin, J.-P. 1994. Algorithm for the rigorous coupled-wave analysis of grating diffraction, *J. Opt. Soc. Am. A*, 11:1321–1331. DOI: 10.1364/josaa.11.001321. 91

[7] Wang, F. and Lakhtakia, A. 2004. Lateral shifts of optical beams on reflection by slanted chiral sculptured thin films, *Opt. Commun.*, 235:107–132. DOI: 10.1016/j.optcom.2004.02.050. 91

[8] Onishi, M., Crabtree, K., and Chipman, R. A. 2011. Formulation of rigorous coupled-wave theory for gratings in bianisotropic media, *J. Opt. Soc. Am. A*, 28:1747–1758. DOI: 10.1364/josaa.28.001747. 91

[9] Polo, J. A., Jr., Mackay, T. G., and Lakhtakia, A. 2013. *Electromagnetic Surface Waves: A Modern Perspective* (Waltham, MA, Elsevier). 91

[10] Waterman, P. C. 1975. Scattering by periodic surfaces, *J. Acoust. Soc. Am.*, 57:791–802 DOI: 10.1121/1.380521. 91

[11] Lakhtakia, A., Varadan, V. K., and Varadan, V. V. 1985. Scattering by a partially illuminated, doubly periodic, doubly infinite surface, *J. Acoust. Soc. Am.*, 77:1999–2004. DOI: 10.1121/1.2022331. 91

[12] Ward, L. 2000. *The Optical Constants of Bulk Materials and Films*, 2nd ed. (Bristol, UK, Institute of Physics). 91

[13] Hodgkinson, I. J. and Wu, Q. h. 1997. *Birefringent Thin Films and Polarizing Elements* (Singapore: World Scientific). DOI: 10.1142/3324. 91

[14] MacLeod, H. A. 2001. *Thin-Film Optical Filters*, 3rd ed. (Boca Raton, FL, CRC Press). DOI: 10.1201/9781420033236. 91

[15] Baumeister, P. W. 2004. *Optical Coating Technology* (Bellingham, WA, SPIE Press). DOI: 10.1117/3.548071. 91

[16] Boardman, A. D., Ed., 1982. *Electromagnetic Surface Modes* (Chicester, UK, Wiley). 91

[17] Turbadar, T. 1959. Complete absorption of light by thin metal films, *Proc. Phys. Soc.*, 73:40–44. DOI: 10.1088/0370-1328/73/1/307. 91

[18] Turbadar, T. 1964. Complete absorption of plane polarized light by thin metal films, *Opt. Acta*, 11:207–210. DOI: 10.1080/713817875. 91

[19] Kretschmann, E. and Raether, H. 1968. Radiative decay of nonradiative surface plasmons excited by light, *Z. Naturforsch. A*, 23:2135–2136. DOI: 10.1515/zna-1968-1247. 91

[20] Otto, A. 1968. Excitation of nonradiative surface plasma waves in silver by the method of frustrated total reflection, *Z. Phys.*, 216:398–410. DOI: 10.1007/bf01391532. 91

[21] Pulsifer, D. P., Faryad, M., and Lakhtakia, A. 2012. Grating-coupled excitation of Tamm waves, *J. Opt. Soc. Am. B*, 29:2260–2269. DOI: 10.1364/josab.29.002260. 91

[22] Faryad, M. and Lakhtakia, A. 2011. Multiple trains of same-color surface plasmon-polaritons guided by the planar interface of a metal and a sculptured nematic thin film. Part V: Grating-coupled excitation, *J. Nanophoton.*, 5:053527. DOI: 10.1117/1.3663210. 91

[23] Kamran, M. and Faryad, M. 2019. Plasmonic sensor using a combination of grating and prism couplings, *Plasmonics*, 14:791–798. DOI: 10.1007/s11468-018-0859-3. 91

[24] Parratt, L. G. 1954. Surface studies of solids by total reflection of X-rays, *Phys. Rev.*, 95:359–369. DOI: 10.1103/physrev.95.359. 91

[25] Mackay, T. G. and Lakhtakia, A. 2019. *Electromagnetic Anisotropy and Bianisotropy: A Field Guide*, 2nd ed. (Singapore: World Scientific). DOI: 10.1142/11351. 91

[26] Kong, J. A. 1972. Theorems of bianisotropic media, *Proc. IEEE*, 60:1036–1046. DOI: 10.1109/proc.1972.8851. 91

[27] Mackay, T. G. and Lakhtakia, A. 2004. Plane waves with negative phase velocity in Faraday chiral mediums, *Phys. Rev. E*, 69:026602. DOI: 10.1103/physreve.69.026602. 91

[28] Mackay, T. G. and Lakhtakia, A. 2009. Negative refraction, negative phase velocity and counterposition in bianisotropic materials and metamaterials, *Phys. Rev. B*, 79:235121. DOI: 10.1103/physrevb.79.235121. 91

[29] Mackay, T. G. and Lakhtakia, A. 2015. *Modern Analytical Electromagnetic Homogenization* (San Rafael, CA, Morgan & Claypool), IOP Concise Physics. DOI: 10.1088/978-1-6270-5427-0. 92

[30] Martín–Palma, R. J. and Lakhtakia, A. 2010. *Nanotechnology: A Crash Course* (Bellingham, WA, SPIE Press). DOI: 10.1117/3.853406. 92

[31] Martín–Palma, R. J. and Martínez-Duart, J. M. 2017. *Nanotechnology for Microelectronics and Photonics*, 2nd ed. (Amsterdam, The Netherlands, Elsevier). 92

[32] Werner, D. H., Ed., 2017. *Broadband Metamaterials in Electromagnetics: Technology and Applications* (Singapore, Pan Stanford). DOI: 10.1201/9781315364438. 92

[33] Sakoda, K. Ed., 2019. *Electromagnetic Metamaterials: Modern Insights into Macroscopic Electromagnetic Fields* (Singapore, Springer). DOI: 10.1007/978-981-13-8649-7. 92

[34] Lakhtakia, A. 1996. Exact analytic solution for oblique propagation in a piezoelectric, continuously twisted, structurally chiral medium, *Appl. Acoust.*, 49:225–236. DOI: 10.1016/s0003-682x(96)00022-9. 92

[35] Lakhtakia, A., Robbie, K., and Brett, M. J. 1997. Spectral Green's function for wave excitation and propagation in a piezoelectric, continuously twisted, structurally chiral medium, *J. Acoust. Soc. Am.*, 101:2052–2058. DOI: 10.1121/1.418137. 92

APPENDIX A

3 × 3 Dyadics

A 3×3 dyad is composed of two 3-vectors:

$$\underline{d} = \underline{a}\,\underline{b}. \tag{A.1}$$

While $\underline{d} \cdot \underline{p} = \underline{a}\,(\underline{b} \cdot \underline{p})$ and $\underline{p} \cdot \underline{d} = (\underline{p} \cdot \underline{a})\,\underline{b}$ are vectors, $\underline{d} \times \underline{p} = \underline{a}(\underline{b} \times \underline{p})$ and $\underline{p} \times \underline{d} = (\underline{p} \times \underline{a})\underline{b}$ are dyads. The transpose of a dyad $\underline{a}\,\underline{b}$ is the dyad $\underline{b}\,\underline{a}$. Sometimes, dyads are called bivectors. Some authors prefer to write the dyad $\underline{a}\,\underline{b}$ as $\underline{b} \otimes \underline{a}$.

A 3×3 dyadic $\underline{\underline{M}}$ is a linear mapping from $\underline{a}$ to $\underline{b}$:

$$\underline{b} = \underline{\underline{M}} \cdot \underline{a}. \tag{A.2}$$

The 3×3 identity dyadic $\underline{\underline{I}}$ is such that $\underline{a} \cdot \underline{\underline{I}} = \underline{\underline{I}} \cdot \underline{a} = \underline{a}$; likewise, the 3×3 null dyadic $\underline{\underline{0}}$ is defined so that $\underline{a} \cdot \underline{\underline{0}} = \underline{\underline{0}} \cdot \underline{a}$ equals the null vector $\underline{0}$. The general representation of a dyadic is as a sum of dyads; i.e.,

$$\underline{\underline{M}} = \sum_{\ell=1,2,\ldots} \underline{\underline{d}}_\ell = \sum_{\ell=1,2,\ldots} M_\ell\,\hat{\underline{u}}_\ell\hat{\underline{v}}_\ell, \tag{A.3}$$

where M_ℓ are some scalar coefficients, while $\hat{\underline{u}}_\ell$ and $\hat{\underline{v}}_\ell$ are vectors of unit magnitude. Thus, every dyad is a dyadic, but not every dyadic is a dyad.

All 3-vectors can be written using matrix notation. The vector $\underline{a} = \underline{u}_x\,a_x + \underline{u}_y\,a_y + \underline{u}_z\,a_z$ in a Cartesian coordinate system is equivalent to the column 3-vector

$$\underline{a} \equiv \begin{bmatrix} a_x \\ a_y \\ a_z \end{bmatrix}, \tag{A.4}$$

and likewise for the vector $\underline{b}$. Thus, the dyad $\underline{d} = \underline{a}\,\underline{b}$ is equivalent to the 3×3 matrix

$$\underline{d} \equiv \begin{bmatrix} a_x b_x & a_x b_y & a_x b_z \\ a_y b_x & a_y b_y & a_y b_z \\ a_z b_x & a_z b_y & a_z b_z \end{bmatrix}. \tag{A.5}$$

Hence, dyadics in electromagnetics can be written as 3×3 matrixes. The identity dyadic $\underline{\underline{I}}$ is equivalent to the 3×3 identity matrix, and the null dyadic $\underline{\underline{0}}$ to the 3×3 null matrix. The usual algebra of matrixes can thus be used for dyadics as well.

The trace of a dyadic is the sum of the diagonal elements in its matrix representation. Likewise, the determinant of a dyadic is the same as the determinant of its equivalent matrix. A dyadic can be transposed in the same way as a matrix. A dyadic can be inverted if it is nonsingular (i.e., if its determinant is not equal to zero).

The antisymmetric dyadic

$$\underline{a} \times \underline{\underline{I}} = \underline{\underline{I}} \times \underline{a} \tag{A.6}$$

$$= a_x \left(\underline{u}_z \underline{u}_y - \underline{u}_y \underline{u}_z \right) + a_y \left(\underline{u}_x \underline{u}_z - \underline{u}_z \underline{u}_x \right) + a_z \left(\underline{u}_y \underline{u}_x - \underline{u}_x \underline{u}_y \right) \tag{A.7}$$

$$\equiv \begin{bmatrix} 0 & -a_z & a_y \\ a_z & 0 & -a_x \\ -a_y & a_x & 0 \end{bmatrix} \tag{A.8}$$

is often useful to denote gyrotropic constitutive properties that are characteristic of ferrites and plasmas. The simplest antisymmetric dyadic is $\hat{\underline{u}} \times \underline{\underline{I}}$, where $\hat{\underline{u}}$ is any vector of unit magnitude. The trace of any antisymmetric dyadic is zero.

Even vector differential operators can be thought of as 3 × 3 dyadics. Thus, the *curl* operator is written as $\nabla \times \underline{\underline{I}}$ and the *divergence* operator as $\nabla \cdot \underline{\underline{I}}$, with

$$\nabla = \underline{u}_x \frac{\partial}{\partial x} + \underline{u}_y \frac{\partial}{\partial y} + \underline{u}_z \frac{\partial}{\partial z} \tag{A.9}$$

written as a vector in a Cartesian coordinate system.

APPENDIX B

Mathematica™ Codes

B.1 BIANISOTROPIC MULTILAYERED SLAB WITH PLANAR INTERFACES

The Mathematica™ code used to generate the numerical data plotted in Fig. 3.2 follows. The code used to generate the numerical data plotted in Fig. 5.1 is a straightforward specialization of that code.

(* Inputs *)

$\lambda0 = 560 \times 10^{-9}; \epsilon0 = 8.8541878 \times 10^{-12}; \mu0 = 4\pi \times 10^{-7}; \eta0 = \sqrt{\frac{\mu0}{\epsilon0}}; k0 = \frac{2\pi}{\lambda0};$

$\omega = \frac{k0}{\sqrt{\epsilon0\,\mu0}};$

$dA = 0.9\,\lambda0; dB = 1.8\,\lambda0; dC = 1.2\,\lambda0; n1 = 1; n2 = 1;$

zeromat=0 IdentityMatrix[3];

$\epsilon A = \epsilon0\ \text{DiagonalMatrix}[\{2.9 + 0.04\,I, 2.5 + 0.03\,I, 2.1 + 0.02\,I\}];$

$\xi A = I\ \sqrt{\epsilon0\,\mu0}\ \text{DiagonalMatrix}[\{0.10 + 0.0008\,I, 0.07 + 0.0006\,I, 0.06 + 0.0005\,I\}];$

$\zeta A = -\xi A;\ \mu A = \mu0\ \text{DiagonalMatrix}[\{1.10 + 0.025\,I, 1.05 + 0.015\,I, 1.02 + 0.012\,I\}];$

$\epsilon B = \epsilon0\ \text{DiagonalMatrix}[\{2.4 + 0.03\,I, 1.9 + 0.02\,I, 2.4 + 0.03\,I\}];$

$\xi B = \text{zeromat};\ \zeta B = \text{zeromat};\ \mu B = \mu0\ \text{IdentityMatrix}[3];$

$\epsilon C = \epsilon0\ \text{DiagonalMatrix}[\{4.1 + 0.08\,I, 4.1 + 0.08\,I, 3.4 + 0.06\,I\}];$

$\epsilon C[[1, 2]] = -I\ \epsilon0\ (0.8 + 0.005\,I);\ \epsilon C[[2, 1]] = -\epsilon C[[1, 2]];$

$\xi C = I\ \sqrt{\epsilon0\,\mu0}\ \text{DiagonalMatrix}[\{0.06 + 0.0012\,I, 0.06 + 0.0012\,I, 0.04 + 0.001\,I\}];$

$\xi C[[1, 2]] = \sqrt{\epsilon0\,\mu0}\ (0.01 + 0.0003\,I);\ \xi C[[2, 1]] = -\xi C[[1, 2]];\ \zeta C = -\xi C;$

μC = μ0 DiagonalMatrix[{1.3 + 0.04 I, 1.3 + 0.04 I, 1.1 + 0.03 I}];

μC[[1, 2]] = $-I$ μ0 (0.35 + 0.003 I); μC[[2, 1]] = $-\mu$C[[1, 2]];

(* Initialization: for loops over θ_{inc} and ψ *)
θcountmax = 36; ψcountmax = 36; forcount = 0;

plotAs = Array[0&, {(θcountmax + 1) (ψcountmax + 1), 3}];

plotAp = Array[0&, {(θcountmax + 1) (ψcountmax + 1), 3}];

plotRss = Array[0&, {(θcountmax + 1) (ψcountmax + 1), 3}];

plotRsp = Array[0&, {(θcountmax + 1) (ψcountmax + 1), 3}];

plotRps = Array[0&, {(θcountmax + 1) (ψcountmax + 1), 3}];

plotRpp = Array[0&, {(θcountmax + 1) (ψcountmax + 1), 3}];

plotTss = Array[0&, {(θcountmax + 1) (ψcountmax + 1), 3}];

plotTsp = Array[0&, {(θcountmax + 1) (ψcountmax + 1), 3}];

plotTps = Array[0&, {(θcountmax + 1) (ψcountmax + 1), 3}];

plotTpp = Array[0&, {(θcountmax + 1) (ψcountmax + 1), 3}];

(* For loops over θ_{inc} and ψ *)
For[θcount = 0, θcount $\leq$ θcountmax, θcount++, θinc = θcount 2.5 Degree;

For[ψcount = 0, ψcount $\leq$ ψcountmax, ψcount++, ψ = ψcount 5 Degree;

forcount = forcount + 1;

If[θcount==θcountmax, θinc = 0.999 N[θinc] Degree];

q = k0 n1 Sin[θinc]; θtr = ArcSin[(n1/n2) Sin[θinc]];

(* Calculate matrixes $\left[\underline{\underline{P}}\right]^{\mathcal{A},\mathcal{B},\mathcal{C}}$ and $\left[\underline{\underline{M}}\right]^{\mathcal{A},\mathcal{B},\mathcal{C}}$ *)

Pmat[ϵ_, ξ_, ζ_, μ_]:=

Module[{eeνzx, eeνzy, ehνzx, ehνzy, heνzx, heνzy, hhνzx, hhνzy, Jmat, m1, m2, m3},

$$ee\nu zx = -\frac{\mu[[3,3]]\,\epsilon[[3,1]]-\xi[[3,3]]\,(\zeta[[3,1]]+(q\,\mathrm{Sin}[\psi]/\omega))}{\mu[[3,3]]\,\epsilon[[3,3]]-\xi[[3,3]]\,\zeta[[3,3]]};$$

$$ee\nu zy = -\frac{\mu[[3,3]]\,\epsilon[[3,2]]-\xi[[3,3]]\,(\zeta[[3,2]]-(q\,\mathrm{Cos}[\psi]/\omega))}{\mu[[3,3]]\,\epsilon[[3,3]]-\xi[[3,3]]\,\zeta[[3,3]]};$$

$$eh\nu zx = \frac{\mu[[3,1]]\,\xi[[3,3]]-\mu[[3,3]]\,(\xi[[3,1]]-(q\,\mathrm{Sin}[\psi]/\omega))}{\mu[[3,3]]\,\epsilon[[3,3]]-\xi[[3,3]]\,\zeta[[3,3]]};$$

$$eh\nu zy = \frac{\mu[[3,2]]\,\xi[[3,3]]-\mu[[3,3]]\,(\xi[[3,2]]+(q\,\mathrm{Cos}[\psi]/\omega))}{\mu[[3,3]]\,\epsilon[[3,3]]-\xi[[3,3]]\,\zeta[[3,3]]};$$

$$he\nu zx = \frac{\zeta[[3,3]]\,\epsilon[[3,1]]-\epsilon[[3,3]]\,(\zeta[[3,1]]+(q\,\mathrm{Sin}[\psi]/\omega))}{\mu[[3,3]]\,\epsilon[[3,3]]-\xi[[3,3]]\,\zeta[[3,3]]};$$

$$he\nu zy = \frac{\zeta[[3,3]]\,\epsilon[[3,2]]-\epsilon[[3,3]]\,(\zeta[[3,2]]-(q\,\mathrm{Cos}[\psi]/\omega))}{\mu[[3,3]]\,\epsilon[[3,3]]-\xi[[3,3]]\,\zeta[[3,3]]};$$

$$hh\nu zx = -\frac{\mu[[3,1]]\,\epsilon[[3,3]]-\zeta[[3,3]]\,(\xi[[3,1]]-(q\,\mathrm{Sin}[\psi]/\omega))}{\mu[[3,3]]\,\epsilon[[3,3]]-\xi[[3,3]]\,\zeta[[3,3]]};$$

$$hh\nu zy = -\frac{\mu[[3,2]]\,\epsilon[[3,3]]-\zeta[[3,3]]\,(\xi[[3,2]]+(q\,\mathrm{Cos}[\psi]/\omega))}{\mu[[3,3]]\,\epsilon[[3,3]]-\xi[[3,3]]\,\zeta[[3,3]]};$$

Jmat = IdentityMatrix[4];

For[$i = 1, i \leq 4, i{+}{+}$, For[$j = 1, j \leq 4, j{+}{+}$, Jmat[[i, j]] = 1];];

m1 = {{ζ[[2, 1]], ζ[[2, 2]], μ[[2, 1]], μ[[2, 2]]}, {$-\zeta$[[1, 1]], $-\zeta$[[1, 2]], $-\mu$[[1, 1]], $-\mu$[[1, 2]]},

{$-\epsilon$[[2, 1]], $-\epsilon$[[2, 2]], $-\xi$[[2, 1]], $-\xi$[[2, 2]]}, {ϵ[[1, 1]], ϵ[[1, 2]], ξ[[1, 1]], ξ[[1, 2]]}};

m2 = DiagonalMatrix[{ζ[[2, 3]] + (q Cos[ψ]/ω), $-\zeta$[[1, 3]] + (q Sin[ψ]/ω),

$-\epsilon$[[2, 3]], ϵ[[1, 3]]}].

Jmat.DiagonalMatrix[{eeνzx, eeνzy, ehνzx, ehνzy}];

m3 = DiagonalMatrix[{μ[[2, 3]], $-\mu$[[1, 3]], $-\xi$[[2, 3]] + (q Cos[ψ]/ω), ξ[[1, 3]]

+(q Sin[ψ]/ω)}].

Jmat.DiagonalMatrix[{heνzx, heνzy, hhνzx, hhνzy}];

ω (m1 + m2 + m3)];

PA = Pmat[ϵA, ξA, ζA, μA]; PB = Pmat[ϵB, ξB, ζB, μB]; PC = Pmat[ϵC, ξC, ζC, μC];

MA = MatrixExp[I dA PA]; MB = MatrixExp[I dB PB]; MC = MatrixExp[I dC PC];

(* Calculate matrixes $\left[\underline{\underline{K}}_{inc,tr} \right]$ *)
 Kmat[θ_, n_]:={{$-$Sin[ψ], $-$Cos[ψ] Cos[θ], $-$Sin[ψ], Cos[ψ] Cos[θ]},

{Cos[ψ], $-$Sin[ψ] Cos[θ], Cos[ψ], Sin[ψ] Cos[θ]},

{$-(n/\eta 0)$ Cos[ψ] Cos[θ], $(n/\eta 0)$ Sin[ψ], $(n/\eta 0)$ Cos[ψ] Cos[θ], $(n/\eta 0)$ Sin[ψ]},

{$-(n/\eta 0)$ Sin[ψ] Cos[θ], $-(n/\eta 0)$ Cos[ψ], $(n/\eta 0)$ Sin[ψ] Cos[θ], $-(n/\eta 0)$ Cos[ψ]}};

Kinc = Kmat[θinc, n1]; Ktr = Kmat[θtr, n2];

(* Calculate reflectances, transmittances, and absorptances *)
rt = Solve[{{ts}, {tp}, {0}, {0}} ==

Inverse[Ktr].MC.MB.MA.Kinc.{{as}, {ap}, {rs}, {rp}}, {rs, rp, ts, tp}];

rss = rt[[1, 1, 2]]/.{as $\rightarrow$ 1, ap $\rightarrow$ 0}; rsp = rt[[1, 1, 2]]/.{as $\rightarrow$ 0, ap $\rightarrow$ 1};

rpp = rt[[1, 2, 2]]/.{ap $\rightarrow$ 1, as $\rightarrow$ 0}; rps = rt[[1, 2, 2]]/.{ap $\rightarrow$ 0, as $\rightarrow$ 1};

tss = rt[[1, 3, 2]]/.{as $\rightarrow$ 1, ap $\rightarrow$ 0}; tsp = rt[[1, 3, 2]]/.{as $\rightarrow$ 0, ap $\rightarrow$ 1};

tpp = rt[[1, 4, 2]]/.{ap $\rightarrow$ 1, as $\rightarrow$ 0}; tps = rt[[1, 4, 2]]/.{ap $\rightarrow$ 0, as $\rightarrow$ 1};

Rss = Abs[rss]^2; Rsp = Abs[rsp]^2; Rps = Abs[rps]^2; Rpp = Abs[rpp]^2;

Tss = $\frac{\text{n2 Re[Cos[}\theta\text{tr]]}}{\text{n1 Cos[}\theta\text{tr]}}$ Abs[tss]^2; Tsp = $\frac{\text{n2 Re[Cos[}\theta\text{tr]]}}{\text{n1 Cos[}\theta\text{tr]}}$ Abs[tsp]^2;

Tps = $\frac{\text{n2 Re[Cos[}\theta\text{tr]]}}{\text{n1 Cos[}\theta\text{tr]}}$ Abs[tps]^2; Tpp = $\frac{\text{n2 Re[Cos[}\theta\text{tr]]}}{\text{n1 Cos[}\theta\text{tr]}}$ Abs[tpp]^2;

As = 1 $-$ Rss $-$ Rps $-$ Tss $-$ Tps;

Ap = 1 – Rpp – Rsp – Tpp – Tsp;

(* Store reflectances, transmittances, and absorptances in arrays for plotting *)
plotAs[[forcount, 1]] = $N[\theta\text{inc}]$/Degree; plotAs[[forcount, 2]] = $N[\psi]$/Degree;

plotAs[[forcount, 3]] = As;

plotAp[[forcount, 1]] = $N[\theta\text{inc}]$/Degree; plotAp[[forcount, 2]] = $N[\psi]$/Degree;

plotAp[[forcount, 3]] = Ap;

plotRss[[forcount, 1]] = $N[\theta\text{inc}]$/Degree; plotRss[[forcount, 2]] = $N[\psi]$/Degree;

plotRss[[forcount, 3]] = Rss;

plotRsp[[forcount, 1]] = $N[\theta\text{inc}]$/Degree; plotRsp[[forcount, 2]] = $N[\psi]$/Degree;

plotRsp[[forcount, 3]] = Rsp;

plotRps[[forcount, 1]] = $N[\theta\text{inc}]$/Degree; plotRps[[forcount, 2]] = $N[\psi]$/Degree;

plotRps[[forcount, 3]] = Rps;

plotRpp[[forcount, 1]] = $N[\theta\text{inc}]$/Degree; plotRpp[[forcount, 2]] = $N[\psi]$/Degree;

plotRpp[[forcount, 3]] = Rpp;

plotTss[[forcount, 1]] = $N[\theta\text{inc}]$/Degree; plotTss[[forcount, 2]] = $N[\psi]$/Degree;

plotTss[[forcount, 3]] = Tss;

plotTsp[[forcount, 1]] = $N[\theta\text{inc}]$/Degree; plotTsp[[forcount, 2]] = $N[\psi]$/Degree;

plotTsp[[forcount, 3]] = Tsp;

plotTps[[forcount, 1]] = $N[\theta\text{inc}]$/Degree; plotTps[[forcount, 2]] = $N[\psi]$/Degree;

plotTps[[forcount, 3]] = Tps;

plotTpp[[forcount, 1]] = $N[\theta\text{inc}]$/Degree; plotTpp[[forcount, 2]] = $N[\psi]$/Degree;

plotTpp[[forcount, 3]] = Tpp;

]]

B.2   ISOTROPIC DIELECTRIC MULTILAYERED SLAB WITH A PERIODIC INTERFACE

The Mathematica™ code used to generate the numerical data plotted in Fig. 5.2 follows. The code for Fig. 5.3 is the same except that the parameters L_g, M_t, N_{II}, and N_s take different values.

(* Inputs *)

$\lambda 0 = 560 \times 10^{-9}$; $\epsilon 0 = 8.8541878 \times 10^{-12}$; $\mu 0 = 4\pi \times 10^{-7}$; $\eta 0 = \sqrt{\frac{\mu 0}{\epsilon 0}}$; k0 $= \frac{2\pi}{\lambda 0}$;

$\omega = \frac{\text{k0}}{\sqrt{\epsilon 0\, \mu 0}}$; Lg $= \lambda 0/40$; gmin $= -$Lg/2; gmax $=$ Lg/2; Lx $= 0.9\,\lambda 0$; Mt $= 4$;

ϵA $= (2 + 0.01\ I)\ \epsilon 0$; ϵB $= (4 + 0.03\ I)\ \epsilon 0$; ϵC $= (3 + 0.05\ I)\ \epsilon 0$;

dA $= 0.9\,\lambda 0$; dB $= 1.8\,\lambda 0$; dC $= 1.2\,\lambda 0$; d $=$ dA $+$ dB $+$ dC;

NI $= 1$; NII $= 5$; NIII $= 1$; NIV $= 1$; Ns $=$ NI $+$ NII $+$ NIII $+$ NIV;

Zeromat $= 0$ IdentityMatrix[2 Mt $+ 1$]; Zeromat2 $= 0$ IdentityMatrix[2 (2 Mt $+ 1$)];

Zeromat4 $= 0$ IdentityMatrix[4 (2 Mt $+ 1$)];

Zeromat42 $=$ ArrayFlatten[{{Zeromat2}, {Zeromat2}}];

(* Calculate $z_0, z_1, \ldots, z_{N_s}$ *)
zcoord $=$ Array[0&, Ns $+ 1$];

For$\left[i = 2, i \leq \text{Ns} - 1, i\text{++}, \text{zcoord}[[i]] = \text{dA} + \text{gmin} + \frac{(i-2)\ (\text{gmax}-\text{gmin})}{\text{NII}} \right]$

zcoord[[Ns]] $=$ dA $+$ dB; zcoord[[Ns $+ 1$]] $=$ d;

(* Calculate $\Upsilon^{(m)}$ for $g(x) = \frac{L_g}{2}\sin\left(\frac{2\pi x}{L_x}\right)$ *)

Υ[n_, z_]:=Module[{x, xlower, xupper, ans},

If[z < dA, xupper = Lx + $\frac{\text{Lx}}{2\pi}$ArcSin$\left[\frac{2\,(z-\text{dA})}{\text{Lg}}\right]$; xlower = $\frac{\text{Lx}}{2}$ + (Lx - xupper);

ans = $\frac{1}{\text{Lx}}$ Integrate$\left[\text{Exp}\left[-I\,\frac{2\,\pi\,n\,x}{\text{Lx}}\right], \{x, \text{xlower}, \text{xupper}\}\right]$];

If[z > dA, xlower = $\frac{\text{Lx}}{2\pi}$ArcSin$\left[\frac{2\,(z-\text{dA})}{\text{Lg}}\right]$; xupper = $\frac{\text{Lx}}{2}$ - xlower;

ans = $\frac{1}{\text{Lx}}$ (Integrate$\left[\text{Exp}\left[-I\,\frac{2\,\pi\,n\,x}{\text{Lx}}\right], \{x, 0, \text{xlower}\}\right]$

+ Integrate$\left[\text{Exp}\left[-I\,\frac{2\,\pi\,n\,x}{\text{Lx}}\right], \{x, \text{xupper}, \text{Lx}\}\right])$];

If[z==dA, xlower = $\frac{\text{Lx}}{2}$; xupper = Lx;

ans = $\frac{1}{\text{Lx}}$ Integrate$\left[\text{Exp}\left[-I\,\frac{2\,\pi\,n\,x}{\text{Lx}}\right], \{x, \text{xlower}, \text{xupper}\}\right]$]; ans]

Υmat = Zeromat;

For[i = 1, i < 2Mt + 2, i++, For[j = 1, j < 2Mt + 2, j++,

Υmat[[i, j]] = (i − 1) + (1 − j)]];

Υarray = Array[Zeromat&, NII];

For[k = 1, k ≤ NII, k++, For[i = 1, i < 2 Mt + 2, i++, For[j = 1, j < 2 Mt + 2, j++,

Υarray[[k]][[i, j]] = Υ[Υmat[[i, j]], (zcoord[[k + 2]] + zcoord[[k + 1]])/2]]]];

(* Calculate $\epsilon^{(m)}$ *)

ϵarray = Array[Zeromat&, Ns];

ϵarray[[1]] = ϵA IdentityMatrix[2 Mt + 1];

For[k = 1, k ≤ NII, k++,

ϵarray[[NI + k]] = ϵA IdentityMatrix[2 Mt + 1] + (ϵB − ϵA) Υarray[[k]]];

ϵarray[[Ns − 1]] = ϵB IdentityMatrix[2 Mt + 1];

```
εarray[[Ns]] = εC IdentityMatrix[2 Mt + 1];
```

```
(* Initialization: for loop over θinc *)
θcountmax = 179;
```

```
plotAs = Array[0&, {θcountmax + 1, 2}]; plotAp = Array[0&, {θcountmax + 1, 2}];
```

```
plotRss0 = Array[0&, {θcountmax + 1, 2}]; plotRpp0 = Array[0&, {θcountmax + 1, 2}];
```

```
plotRss1 = Array[0&, {θcountmax + 1, 2}]; plotRpp1 = Array[0&, {θcountmax + 1, 2}];
```

```
plotTss0 = Array[0&, {θcountmax + 1, 2}]; plotTpp0 = Array[0&, {θcountmax + 1, 2}];
```

```
plotTss1 = Array[0&, {θcountmax + 1, 2}]; plotTpp1 = Array[0&, {θcountmax + 1, 2}];
```

```
(* For loop over θinc *)
For[θcount = 0, θcount ≤ θcountmax, θcount++, θ = θcount 0.5 Degree;
```

$$\kappa[n\_]:=\text{k0 Sin}[\theta] + \frac{2\pi n}{\text{Lx}}; \alpha[n\_]:=\sqrt{\text{k0}^2 - \kappa[n]^2};$$

```
Kmat = Zeromat; αmat = Zeromat;
```

```
For[i = −Mt, i < Mt + 1, i++, Kmat[[i + Mt + 1, i + Mt + 1]] = κ[i]];
```

```
For[i = −Mt, i < Mt + 1, i++, αmat[[i + Mt + 1, i + Mt + 1]] = α[i]];
```

$$(*\text{ Calculate matrixes } \left[\underline{\underline{\breve{P}}}\right]^{(m)} *)$$
```
Pmat[εmat_]:=Module[{Pa, Pb, m1, m2},
```

```
Pa = ArrayFlatten[{{Zeromat, Zeromat, Zeromat, μ0 IdentityMatrix[2 Mt + 1]},
```

```
{Zeromat, Zeromat, −μ0 IdentityMatrix[2 Mt + 1], Zeromat},
```

```
{Zeromat, −εmat, Zeromat, Zeromat},{εmat, Zeromat, Zeromat, Zeromat}}];
```

$$\text{m1} = -(\text{Kmat.Inverse}[\varepsilon\text{mat}].\text{Kmat}); \text{m2} = \frac{1}{\mu 0} (\text{Kmat.Kmat});$$

```
Pb = ArrayFlatten[{{Zeromat, Zeromat, Zeromat, −m1},
```

{Zeromat, Zeromat, Zeromat, Zeromat}, {Zeromat, m2, Zeromat, Zeromat},

{Zeromat, Zeromat, Zeromat, Zeromat}}];

$(\omega \, \text{Pa}) + \frac{1}{\omega} \, \text{Pb}\big]$;

Parray = Array[Zeromat4&, Ns];

For[$k = 1, k \leq$ Ns, k++, Parray[[k]] = Pmat[ϵarray[[k]]]];

(* Diagonalization of matrixes $\left[\underline{\underline{\breve{P}}}\right]^{(m)}$: calculation of $\left[\underline{\underline{\breve{G}}}\right]^{(m)}$ and $\left[\underline{\underline{\breve{V}}}\right]^{(m)}$ *)
DiagPmat[P_]:=Module[{Pevals, Pevecs, Vmat, etemp, vtemp, Gmat},

{Pevals, Pevecs} = Eigensystem[P];

Vmat = Transpose[Pevecs];

For[$i = 1, i < 4\,(2\,\text{Mt} + 1), i$++, For[$j = 1, j < 4\,(2\,\text{Mt} + 1) + 1 - i, j$++,

If[Im[Pevals[[j]]] < Im[Pevals[[$j + 1$]]],

etemp = Pevals[[j]]; vtemp = Vmat[[All, j]];

Pevals[[j]] = Pevals[[$j + 1$]]; Pevals[[$j + 1$]] = etemp;

Vmat[[All, j]] = Vmat[[All, $j + 1$]]; Vmat[[All, $j + 1$]] = vtemp;]]];

Gmat = DiagonalMatrix[Pevals];

{Gmat, Vmat}];

Garray = Array[Zeromat4&, Ns]; Varray = Array[Zeromat4&, Ns];

For[$k = 1, k \leq$ Ns, k++, {Garray[[k]], Varray[[k]]} = DiagPmat[Parray[[k]]]];

Gupperarray = Array[Zeromat2&, Ns]; Glowerarray = Array[Zeromat2&, Ns];

```
For[k = 1, k ≤ Ns, k++,

For[i = 1, i ≤ 2 (2 Mt + 1), i++, Gupperarray[[k]][[i, i]] = Garray[[k]][[i, i]];

Glowerarray[[k]][[i, i]] = Garray[[k]][[2 (2 Mt + 1) + i, 2 (2 Mt + 1) + i]]]];
```

(* Calculation of matrixes $\left[\underset{=e,h}{\breve{Y}}^{\text{inc,ref}}\right]$ *)

```
eYinc = ArrayFlatten [{{Zeromat, -1/k0 αmat} , {IdentityMatrix[2 Mt + 1], Zeromat}}] ;

hYinc = -1/η0 ArrayFlatten [{{1/k0 αmat, Zeromat} , {Zeromat, IdentityMatrix[2 Mt + 1]}}] ;

eYref = ArrayFlatten [{{Zeromat, 1/k0 αmat} , {IdentityMatrix[2 Mt + 1], Zeromat}}] ;

hYref = 1/η0 ArrayFlatten [{{1/k0 αmat, Zeromat} , {Zeromat, -IdentityMatrix[2 Mt + 1]}}] ;

eYtr = eYinc; hYtr = hYinc;
```

(* Calculation of matrix $\left[\underset{=}{\breve{Z}}\right]^{(0)}$ *)

```
Zarray = Array[Zeromat42&, Ns];

Zarray[[Ns]] = ArrayFlatten[{{eYtr}, {hYtr}}];

GetZmat[Z_, V_, Glower_, Gupper_, zupper_, zlower_]:=

Module[{Xmat, Xupper, Xlower, Umat},
Xmat = Inverse[V].Z;

Xupper = Zeromat2; Xlower = Zeromat2;

For[i = 1, i ≤ 2 (2 Mt + 1), i++, For[j = 1, j ≤ 2 (2 Mt + 1), j++,

Xupper[[i, j]] = Xmat[[i, j]]; Xlower[[i, j]] = Xmat[[2 (2 Mt + 1) + i, j]];]];

Umat = MatrixExp[-I (zupper − zlower) Glower].Xlower.Inverse[Xupper].

MatrixExp[I (zupper − zlower) Gupper];
```

V.ArrayFlatten[{{IdentityMatrix[2 (2 Mt + 1)]}, {Umat}}]];

For[k = Ns − 1, $k \geq 1$, k−,

Zarray[[k]] = GetZmat[Zarray[[k + 1]], Varray[[k + 1]],

Glowerarray[[k + 1]], Gupperarray[[k + 1]], zcoord[[k + 2]], zcoord[[k + 1]]]];

Z0 = GetZmat[Zarray[[1]], Varray[[1]], Glowerarray[[1]],

Gupperarray[[1]], zcoord[[2]], zcoord[[1]]];
Z0upper = Zeromat2; Z0lower = Zeromat2;

For[i = 1, $i \leq 2$ (2 Mt + 1), i++, For[j = 1, $j \leq 2$ (2 Mt + 1), j++,

Z0upper[[i, j]] = Z0[[i, j]]; Z0lower[[i, j]] = Z0[[2 (2 Mt + 1) + i, j]];];

(* Calculation of vector $\left[\breve{A}\right]$ *)
Avec = Array[0&, 2 (2 Mt + 1)]; Avec[[Mt + 1]] = as;

Avec[[(2 Mt + 1) + Mt + 1]] = ap;

sAvec = Avec/.{as → 1, ap → 0}; pAvec = Avec/.{as → 0, ap → 1};

(* Calculation of vector $\left[\breve{T}\right]$ *)
T0Rvec = Inverse[ArrayFlatten[{{Z0upper, −eYref}, {Z0lower, −hYref}}]].

ArrayFlatten[{{eYinc}, {hYinc}}].Avec;
T0 = Array[0&, 2 (2 Mt + 1)];

For[i = 1, $i \leq 2$ (2 Mt + 1), i++, T0[[i]] = T0Rvec[[i]]]];
sT0 = T0/.{as → 1, ap → 0}; pT0 = T0/.{as → 0, ap → 1};

Rvec = Array[0&, 2 (2 Mt + 1)];

For[i = 1, $i \leq 2$ (2 Mt + 1), i++, Rvec[[i]] = T0Rvec[[2 (2 Mt + 1) + i]]]];

sRvec = Rvec/.{as → 1, ap → 0}; pRvec = Rvec/.{as → 0, ap → 1};

```
GetTvec[T_, V_, Z_, Gupper_, zupper_, zlower_]:=Module[{Xmat, Xupper},

Xmat = Inverse[V].Z;

Xupper = Zeromat2;

For[i = 1, i ≤ 2 (2 Mt + 1), i++, For[j = 1, j ≤ 2 (2 Mt + 1), j++,

Xupper[[i, j]] = Xmat[[i, j]]];];

Inverse[Xupper].MatrixExp[I (zupper − zlower) Gupper].T];

sTarray = Array[0&, 2 (2 Mt + 1)];

For[k = 1, k ≤ Ns, k++,

If[k == 1, sTarray[[k]] = GetTvec[sT0, Varray[[k]], Zarray[[k]], Gupperarray[[k]],

zcoord[[k + 1]], zcoord[[k]]],
sTarray[[k]] = GetTvec[sTarray[[k − 1]], Varray[[k]], Zarray[[k]], Gupperarray[[k]],

zcoord[[k + 1]], zcoord[[k]]]]];
sTvec = sTarray[[Ns]];

pTarray = Array[0&, 2 (2 Mt + 1)];

For[k = 1, k ≤ Ns, k++,

If[k == 1, pTarray[[k]] = GetTvec[pT0, Varray[[k]], Zarray[[k]], Gupperarray[[k]],

zcoord[[k + 1]], zcoord[[k]]],
pTarray[[k]] = GetTvec[pTarray[[k − 1]], Varray[[k]], Zarray[[k]], Gupperarray[[k]],

zcoord[[k + 1]], zcoord[[k]]]]];
pTvec = pTarray[[Ns]];
```

(* Calculation of reflectances, transmittances, and absorptances *)
rss[n_]:=sRvec[[Mt + 1 + n]]; rpp[n_]:=pRvec[[(2 Mt + 1) + Mt + 1 + n]];

tss[n_]:=sTvec[[Mt + 1 + n]]; tpp[n_]:=pTvec[[(2Mt + 1) + Mt + 1 + n]];

Rss[n_]:= $\frac{\mathrm{Re}[\alpha[n]]}{\alpha[0]}$ Abs[rss[n]]^2; Rpp[n_]:= $\frac{\mathrm{Re}[\alpha[n]]}{\alpha[0]}$ Abs[rpp[n]]^2;

Tss[n_]:= $\frac{\mathrm{Re}[\alpha[n]]}{\alpha[0]}$ Abs[tss[n]]^2; Tpp[n_]:= $\frac{\mathrm{Re}[\alpha[n]]}{\alpha[0]}$ Abs[tpp[n]]^2;

As = 1 − Sum[Rss[n] + Tss[n], {n, −Mt, Mt}];

Ap = 1 − Sum[Rpp[n] + Tpp[n], {n, −Mt, Mt}];

(* Store reflectances, transmittances, and absorptances in arrays for plotting *)
plotAs[[θcount + 1, 1]] = N[θ]/Degree; plotAs[[θcount + 1, 2]] = As;

plotAp[[θcount + 1, 1]] = N[θ]/Degree; plotAp[[θcount + 1, 2]] = Ap;

plotRss0[[θcount + 1, 1]] = N[θ]/Degree; plotRss0[[θcount + 1, 2]] = Rss[0];

plotRpp0[[θcount + 1, 1]] = N[θ]/Degree; plotRpp0[[θcount + 1, 2]] = Rpp[0];

plotRss1[[θcount + 1, 1]] = N[θ]/Degree; plotRss1[[θcount + 1, 2]] = Rss[−1];

plotRpp1[[θcount + 1, 1]] = N[θ]/Degree; plotRpp1[[θcount + 1, 2]] = Rpp[−1];

plotTss0[[θcount + 1, 1]] = N[θ]/Degree; plotTss0[[θcount + 1, 2]] = Tss[0];

plotTpp0[[θcount + 1, 1]] = N[θ]/Degree; plotTpp0[[θcount + 1, 2]] = Tpp[0];

plotTss1[[θcount + 1, 1]] = N[θ]/Degree; plotTss1[[θcount + 1, 2]] = Tss[−1];

plotTpp1[[θcount + 1, 1]] = N[θ]/Degree; plotTpp1[[θcount + 1, 2]] = Tpp[−1];

]

Authors' Biographies

TOM G. MACKAY

Tom G. Mackay is a graduate of the Universities of Edinburgh, Glasgow, and Strathclyde. His university education was supported by the *British Heart Foundation* and *The Carnegie Trust for The Universities of Scotland*. He is a reader in the School of Mathematics at the University of Edinburgh, and also an adjunct professor in the Department of Engineering Science and Mechanics at The Pennsylvania State University. In 2006/07 he held a *Royal Society of Edinburgh/Scottish Executive Support Research Fellowship* and in 2009/2010 a *Royal Academy of Engineering/Leverhulme Trust Senior Research Fellowship*. He is a Fellow of the Institute of Physics (UK) and SPIE—The International Society for Optics and Photonics. He has been working on the electromagnetic theory of complex mediums, including homogenization and surface waves, for the past 22 years.

AKHLESH LAKHTAKIA

Akhlesh Lakhtakia is Evan Pugh University Professor and Charles Godfrey Binder (Endowed) Professor of Engineering Science and Mechanics at The Pennsylvania State University. He received his B.Tech. (1979) and D.Sc. (2006) degrees in Electronics Engineering from the Institute of Technology, Banaras Hindu University, and his M.S. (1981) and Ph.D. (1983) degrees in Electrical Engineering from the University of Utah. He was the Editor-in-Chief of the *Journal of Nanophotonics* from its inception in 2007 through 2013. He has been elected a Fellow of the American Association for the Advancement of Sciences, American Physical Society, Institute of Physics (UK), Optical Society of America, SPIE—The International Society for Optics and Photonics, Institute of Electrical and Electronics Engineers, Royal Society of Chemistry, and Royal Society of Arts. His current research interests relate to electromagnetic fields in complex mediums, sculptured thin films, surface multiplasmonics and electromagnetic surface waves, mimumes, multicontrollable metasurfaces, thin-film solar cells, forensic science, bioreplication, engineered biomimicry, and biologically inspired design for environment.

Printed in the United States
by Baker & Taylor Publisher Services